AMERICANA LIBRARY

ROBERT E. BURKE, EDITOR

WM. E. SMYTHE.

The
Conquest of Arid America

BY
WILLIAM E. SMYTHE

Introduction by Lawrence B. Lee

UNIVERSITY OF WASHINGTON PRESS

SEATTLE AND LONDON

Copyright © 1899, 1905 by William E. Smythe
University of Washington Press Americana Library edition 1969
Originally published by Harper and Brothers; revised edition
published by the Macmillan Company
"Introduction to the 1969 Edition" by Lawrence B. Lee
copyright © 1969 by the University of Washington Press
Library of Congress Catalog Card Number 5-41786
Printed in the United States of America

TO

MY WIFE

EMANCIPATION

The Nation reaches its hand into the Desert,
And lo! private monopoly in water and in land is scourged
from that holiest of temples,—the place where men
labor and build their homes!

The Nation reaches its hand into the Desert.
The wasting floods stand back, the streams obey their
master, and the stricken forests spring to life again
upon the forsaken mountains!

The Nation reaches its hand into the Desert.
The barred doors of the sleeping empire are flung wide
open to the eager and the willing, that they may
enter in and claim their heritage!

The Nation reaches its hand into the Desert.
That which lay beyond the grasp of the Individual yields
to the hand of Associated Man. Great is the
Achievement,—greater the Prophecy!

FOREWORD

In the summer of 1899, on a remote ranch in the desert country of Northeastern California, twelve miles from the nearest house ("and no one living there," as my wife caustically remarked) I put the finishing touches to the first edition of this book. Now, six years to a day, in my library at San Diego, with its seaward gable commanding a view of a populous town and a wide landscape of bay, ocean, mountains, islands, and the bold promontory of Point Loma, I finish my revision of a new and enlarged edition which brings the story of Arid America down to date.

There is a curious coincidence between the changed scene and conditions of my work at these two periods and the change in the fortunes of the Cause with which I deal. The irrigation movement of 1899 was comparatively remote from the great heart of the Nation, domiciled in distant deserts, a long way from "the nearest house" (and almost "nobody lived there," too), while the irrigation movement of 1905 is planted in the heart of populous towns and intimately related to the commerce of the world.

Six years ago national irrigation was a dream; to-day, the dream has come true. Where the subject interested one man at the time the book was first written, it probably interests at least a thousand men at the time when it is re-written.

FOREWORD

The green fruit has ripened on the bough of time in the sunshine of events. In the name of all my comrades, living and dead, I thank God that this is so! And yet it is those who have lived nearest to this growth of institutions who realize most keenly that we have but crossed the threshold of our new epoch. There is a vast work for us yet to do, there will be a vast work for us to pass on to our sons, and for our sons to pass on to their sons. The inspiring thing is that a point has been reached when the real builders of the world—the men who clear the brush, level the land, plow, plant, cultivate, and reap—can help in a practical way, because their speeches and books and proclamations are written on the face of the imperishable earth.

We are ready for the homeseeker. The fate of our cause is in his hands. If he fail, all the labor of the years comes to naught and the great hope dies in the morning of its promise. But if he succeed, no imagination can set bounds to the achievement upon which we have entered nor picture the civilization which will rise in the waste places of the West.

"Arid America!" exclaimed Joaquin Miller, "we have watered it with our tears!" And so we have. Now we are to build it in toil, in pain, in patience, humbling ourselves in the dust of failure, yet moving ever forward in that pathway of co-operation and brotherhood of which the Newlands Irrigation Law is the most shining guidepost thus far erected by the genius of our statesmanship.

This book is for all optimistic Americans, but especially, it is for those who have the courage of their optim-

FOREWORD

ism—for the homeseekers who, under the leadership of the paternal Nation, are to grapple with the desert, translate its gray barrenness into green fields and gardens, banish its silence with the laughter of children. This is the breed of men who make the Republic possible, who keep the lamp of faith burning through the night of corrupt commercialism, and who bear the Ark of the Covenant to the Promised Land.

In the following pages I have endeavored to show the relation between the earliest settlers in America and the new army which is now moving toward our Western lands; the peculiar environment of the arid region and the influence which it will exert on its civilization; the lessons to be learned from the more notable of the early pioneer settlements in Colorado, Utah, and California; the natural advantages and present development of the great States and Territories between the Missouri River and the Pacific Ocean; the beginning, progress, and triumph of the national irrigation movement; the work of the remarkable corps of young men organized in the United States Reclamation Service; and, finally, the spirit of what is being done by the partnership of God and mankind in finishing one important corner of the world.

In another work, published almost simultaneously with this, I deal with some of the things which ought to be done to enable the Nation to utilize the surplus energies of our growing national family. There is a wide and inviting field for constructive statesmanship to cultivate before we can hope to proceed in the best way with the building of the Unfinished Republic. But something has

FOREWORD

already been done, and it is that which is the subject of the present volume.

I hope the book will be of value to several different kinds of people—to the investor, the tourist, the economist, the legislator, the reader of history and travel, and those interested in American resources and institutions generally—but most of all I hope it will be of some practical use to the men and women who are looking for homes under the blue sky of the West.

W. E. S.

SAN DIEGO, CALIFORNIA.

CONTENTS

	PAGE
INTRODUCTION: ON GOING WEST, YOUNG MAN	xvii
INTRODUCTION TO THE 1969 EDITION BY LAWRENCE B. LEE	xxix

PART FIRST

COLONIAL EXPANSION AT HOME

CHAPTER	
I. GREATNESS BY CONTINENTAL CONQUEST	3
II. THE HOME-BUILDING INSTINCT OF THE AMERICAN PEOPLE	12
III. THE BETTER HALF OF THE UNITED STATES	19
IV. THE BLESSING OF ARIDITY	30
V. THE MIRACLE OF IRRIGATION	41

PART SECOND

REAL UTOPIAS OF THE ARID WEST

I. THE MORMON COMMONWEALTH	51
II. THE GREELEY COLONY OF COLORADO	77
III. THE EVOLUTION OF SOUTHERN CALIFORNIA	92
IV. THE REVOLUTION ON THE PLAINS	106

CONTENTS

PART THIRD

UNDEVELOPED AMERICA

CHAPTER	PAGE
I. THE EMPIRE STATE OF THE PACIFIC	121
II. THE NEW DAY IN COLORADO	161
III. THE PLEASANT LAND OF UTAH	175
IV. THE CRUDE STRENGTH OF IDAHO	185
V. THE GIANT WASHINGTON	197
VI. OREGON IN TRANSITION	205
VII. THE RISING STATE OF NEVADA	213
VIII. THE UNKNOWN LAND OF WYOMING	221
IX. THE PROSPERITY OF MONTANA	232
X. THE AWAKENING OF NEW MEXICO	238
XI. THE BUDDING CIVILIZATION OF ARIZONA	248

PART FOURTH

THE TRIUMPH OF THE MOVEMENT

I. THE RISE OF A NEW CAUSE	261
II. ON THE ANVIL OF CONGRESS	275
III. IRRIGATION IN THE WHITE HOUSE	281
IV. UNCLE SAM'S YOUNG MEN AT WORK	294
V. PREPARING HOMES FOR THE PEOPLE	302
CONCLUSION: MAN'S PARTNERSHIP WITH GOD	327

APPENDIX

I. NOTE AS TO METHODS OF IRRIGATION	333
II. THE NEWLANDS BILL AND THE ACT OF JUNE 17, 1902.	342

INDEX	351

ILLUSTRATIONS

PORTRAIT OF THE AUTHOR.......................*Frontispiece*	
	FACING PAGE
FLOWING WELL, RIVERSIDE CANAL SYSTEM, CALIFORNIA...	12
THE DESERT BEFORE RECLAMATION.......................	24
GROWTH OF VEGETATION IN IMPERIAL VALLEY, CALIFORNIA.	30
PARK AT RIVERSIDE, CALIFORNIA.........................	42
RANCH AT NORTH YAKIMA, WASHINGTON................	46
PALESTINE AND SALT LAKE VALLEY, UTAH................	54
CALIFORNIA CONTRAST, FLOWERS AND SNOW AT PASADENA.	94
THE COLORADO DESERT BEFORE AND AFTER IRRIGATION....	152
VIEWS OF THE NEW TOWN OF IMPERIAL, CALIFORNIA.......	154
IRRIGATED APRICOT ORCHARD, NEAR MONTROSE, COLORADO..	166
VIEW ON GUNNISON RIVER, COLORADO...................	170
MOUTH OF ECHO CANYON, AND WEBER RIVER, UTAH........	180
BUILDING A GOVERNMENT CANAL IN NEVADA...............	214
WHERE THE GATES WERE LIFTED ON THE TRUCKEE RIVER, NEVADA...	216
TWO IRRIGATION STATESMEN AND AN ENGINEER...........	220
EAGLE DAM SITE, RIO GRANDE PROJECT, NEW MEXICO	242
ARIZONA ALFALFA AND BARLEY...........................	250
IRRIGATING A YOUNG ORCHARD IN ARIZONA...............	254
PORTRAIT OF JOHN WESLEY POWELL......................	260
A SAMPLE OF GOVERNMENT WORKS, NEVADA..............	270
PORTRAIT OF FRANCIS G. NEWLANDS......................	276
GOVERNMENT ROAD-BUILDING IN ARIZONA.................	278

ILLUSTRATIONS

	FACING PAGE
PORTRAIT OF THEODORE ROOSEVELT	282
GOVERNMENT CEMENT MILL, ARIZONA	288
PORTRAIT OF ETHAN ALLEN HITCHCOCK	296
PORTRAITS OF CHARLES D. WALCOTT, FREDERICK H. NEWELL, C. E. GRUNSKY, AND GIFFORD PINCHOT	298
SITE OF ROOSEVELT DAM, ARIZONA	308
COLORADO RIVER AT HIGH WATER	310
VIEWS OF KLAMATH PROJECT, OREGON	312
A GOVERNMENT TUNNEL IN NEVADA	316
WORKS OF TRUCKEE-CARSON PROJECT, NEVADA	318

INTRODUCTION

ON GOING WEST, YOUNG MAN

MANY persons ask if I advise them to go West. I invariably reply that it depends upon their temperament. For there is a Western temperament, and there is an Eastern temperament. The person who possesses the former will never be really happy in the East; the person who possesses the latter will never be happy in the West.

The man with the Western temperament loves the unbuilt house and the virgin soil—the vast resources awaiting the conquest of human genius and human labor. He wants to live in a land where things are being done, and where they are to be done yet more in the future. He wants to have a part in doing them,— wants to build the house, plant the ivy, turn rivers out of their courses, drive the desert back inch by inch, carry railroads through unheard-of mountain passes, write constitutions, found cities and states.

The man with the Eastern temperament prefers his civilization ready-made. He loves the old home, the old familiar names and streets, the old associations. He loves the ivy, too, but wants to know that it was planted by his great-grandfather. He wants to feel that cities and states were founded, and constitutions written, by

INTRODUCTION

men who were much wiser than he, because they lived and died so many generations ago. When this man gets West he is homesick. But his brother of the Western temperament works with fierce joy. He is worth ten times as much to himself and to society as he would have been if he had always remained on his native heath.

It is my own feeling (for one is ruled, after all, by one's inherited prejudices and pride of home) that the happiest of all fates is to be born in Massachusetts and to live in California! This is the feeling common to Western men of Eastern antecedents. They vary the localities to suit their birthplace and present residence, but the spirit of the observation remains intact. They are by no means ashamed of the old home. On the contrary, they love and revere it and to revisit it is a delightful experience. But it no longer satisfies their souls, while to stay there long is to hunger for the sight of the great, rugged mountains, for the smell of the desert, for the touch of the unfailing sunshine.

If you feel like that, young man, go West! The Unfinished Republic is calling to you. By all means answer, "Here am I!" You may not be successful in a pecuniary way, yet you will never regret the move. And if you yield to the temptation to go "back East," you will never again be quite satisfied. You will think of the wide landscapes and of the freedom of which they are the sign and symbol, and your heart will quickly traverse the hills and prairies, to nestle in the shadows of the Rockies or the Sierras and abide forever. If it does not, then your temperament is not Western, after all, and you should have lived and died in the land of your ancestors.

INTRODUCTION

A word to the married man: *you* may have the Western temperament, but how about your wife? Her feelings are entitled to consideration, and the time to think of them is before you detach yourself from the old home and its associations. I remember that when I was seeking the assistance of Edward Everett Hale in organizing a new Plymouth, ten years ago, the dear old man turned to his wife and said, with a twinkle in his eye, " Mother, I believe I will go out to Idaho and grow up with the country." She replied, " No, you will stay right here." And, indeed, his own rebuke of the jest was engraved upon the stone above his fireplace:

" Old wood to burn, old books to read, old friends to keep." Dr. Hale is one of the rare souls who can preserve a thoroughly Western temperament while living in an Eastern environment. His influence and fame long since went West and grew up with the country.

But to return to the married man who thinks of moving to improve his condition in life. I hope he will consult the wishes of his " better half." It always seemed to me that the Pilgrim Fathers were glorified at the expense of the Pilgrim Mothers, and I have often wondered how the Mothers felt as they looked upon the receding shores of Old England. The true Western man has no need to ask how the Fathers felt, for he is moved by the same sense of forefatherhood which nerved the arms of the Pilgrims when they went forth to fell the forests and make clearings for their homes.

I venture to say that the average man likes the West. The newness, the bigness, the essential masculinity of the sparsely-peopled wilderness, and of the tasks to which

INTRODUCTION

it invites him, appeal to his pride and strength and kindle his imagination. In a way, he is born again. He turns over a new page in his life history. He does not intend to repeat the old mistakes, and he starts with a fund of knowledge which he did not have when he began his earlier career. He realizes that his relation to the natural resources of the region is like that enjoyed by the men of a century ago in the place whence he came. *They* planted hamlets which grew into cities and thereby enriched their descendants; *he* will do the same. His heart swells as he thinks of his own and his children's future.

But if this is the feeling of the average man who goes West, it is not necessarily shared by the average woman. She is more sensitive to her surroundings. It may be that she has less imagination—that she cannot see the fields and towns which are to be, through the clouds of dust that come swirling from the treeless land. Or it may be—and this is more likely—that she gets the heavier end of the burdens peculiar to the pioneer. "It is easier for a man to get a living in the West, than in the East," one lady said to me, "but women have to scratch harder here than they do back home." It is not difficult to understand how this may be true. The man sees his opportunity in the very fact that he has come to a rich region which is undeveloped, so that he can skim· the cream of its advantages; but the lack of its development deprives the woman of many of the conveniences which she formerly knew. It is undeniable that the household duties falling to the woman's lot, whether she live in the country or the town, are lightened as civilization increases.

INTRODUCTION

Of course, the hardships of pioneering nowadays are not like those of earlier generations, yet it would be quite misleading to say that the women who come to make homes in the new settlements of the West can hope to enjoy all the advantages to which they were accustomed in their Eastern lives.

It is in its natural resources that the West excels, not in its artificial refinements, and the lack of the latter is much more keenly felt by women than by men. Nor is this deficiency seen only in connection with household work. It is felt on the social side of life. It is not that organized society is so different, though naturally the church, the school, the facilities for shopping, and all that attaches to these interests, are in a more primitive state in a new country than in an old one; but it is the fact that the settler has come to a neighborhood not only deficient in numbers and peopled with strangers, but without traditions. I have heard women say, "There is nothing to talk about, and nobody to talk to if there were." This is by no means literally true, yet it conveys an inkling of the truth.

It is only another way of saying that the old friends and old associations mean more to women than to men, or that men are quicker to make new friends and find enjoyment in new associations than are their wives and daughters. Perhaps the reader can now understand why men take more kindly to the West than do their women-folk, or, at least, why the latter require more time to grow into their new environment.

It would be quite misleading to leave the impression that women never like the wilderness. As a matter of

INTRODUCTION

fact, many of them adore it and would not willingly exchange it for the life of large cities or older agricultural communities. As I write, I hear of a young lady who has enjoyed the advantages of Boston and New York, of San Francisco and Los Angeles, but who resists the appeals of her parents to come out of the desert wild where she went for a brief vacation, already lengthened to months, and where she has previously spent weeks which she enthusiastically describes as "the only time I ever really lived." She writes:

"When I am in the city, my happiness depends on people and society, but out here in the deserts and mountains the country itself is satisfying. Perhaps you don't understand what I mean, but I do."

I understand precisely what she means, and so do all men and women who turn their faces toward the clean, beautiful, unpeopled wilderness with the thrill of a lover's heart.

In what has been said I have been thinking chiefly of the millions who will come to conquer the waste places in real pioneer fashion, making settlements where there is now nothing but scattered ranches or great tracts of sage-brush land. These observations by no means apply to life in the larger towns or even in the older agricultural communities. The town life of the West is up to date; the country life, where resources have been quite thoroughly developed, not only up to date, but far ahead of the time when compared with Eastern standards. Take such communities as Riverside and Redlands, in Southern California, for example, and you find social advantages which cannot be matched anywhere else in

INTRODUCTION

the world. In later pages we shall see what these advantages are and how they were attained.

But the man who is going West must remember that he cannot begin where other people ended. That is to say, he cannot have the benefits of pioneering without its drawbacks. If he wants a ready-made civilization, he must pay the price. The men who went to Riverside a generation ago, or to Redlands fifteen years ago, were in at the birth of things. They planted the seed, bore the heat and burden of the day, and reaped the reward. You may be able to buy one of their perfect homes and orchards, and enter immediately into the enjoyment of their social advantages, but you must pay a snug little fortune for the privilege. The alternative is to begin where they began and grind your grist in patience. No way has been discovered by which a man can get something for nothing, even in the glorious West. But for the man who is not doing as well as he ought in the East, the New West is the land of opportunity.

We shall see in subsequent pages the vast margin of undeveloped resources in all our Western States and Territories, and how these resources call loudly for men and capital to come and make use of them. We shall also see how the Nation itself is beginning to do certain things vital to the growth of civilization, yet far beyond the reach of individual enterprise, and how this public policy sets a new star of hope in the sky of our common humanity. Just here we are dealing with individual opportunities, and the most significant of these consists in the fact that *far more may be done with small capital in the West than in the East.*

INTRODUCTION

To begin with, land is cheaper, though richer and more productive. This is the reason that many shrewd farmers are selling their property in the Middle States and beginning life anew between the Rockies and the Sierras. Of the advantages arising from irrigation the reader will get his fill if he persists to the end of my story. For the present let it be said in a word that the man who wants to win an independence from the soil will accomplish more with a given investment of capital, energy, and time in the Irrigated West than anywhere else in the world. He should choose his precise location wisely and it is the object of this book to assist him in doing so; but, in a general way, his true policy is to enlist in the peaceful army which is engaged in the conquest of Arid America.

What has been said of land is true of all other business opportunities. The dollar will go farther in any direction in the undeveloped West than in the older sections of the country. What can a man do with, say, twenty-five thousand dollars, in Boston, New York, Chicago, or even in Eastern cities of the second or third rank? Can he compete with the department stores, the great factories, or the consolidated banks?

In the Far West, one man possessing this sum, or a number of men having it among them, may readily become a considerable financial power. They can even start a national bank, and they will be surprised to find how the deposits flow in from a wide range of sparsely settled country, and surprised again to learn with what security and profit they can loan the capital at their disposal. Many experienced bankers have been turned

INTRODUCTION

out of comfortable berths by bank consolidations in New England and elsewhere during the past few years. Nearly all of them have moderate means or an acquaintance and standing which would enable them to command a reasonable working capital. Instead of being embittered by the change which has overtaken them, they ought to be inspired by the field of usefulness it opens to them. The New West yearns for such men—for their capital, their knowledge, their training, their proved integrity. It will give them a chance to become leaders where they were formerly but followers.

The same is true of merchants and, in less degree, of manufacturers. Business is done on a smaller scale; the country is growing enormously, and this growth will continue indefinitely; new towns and agricultural districts are springing up where silence and desolation have reigned for ages; the region is measurably free from domination by great capitalists. Doubtless the time will come when the same economic forces which closed most of the commercial avenues to small capital in the East will produce the same result in the West, but it has not come yet. There is probably another generation of prosperity for enterprising men who follow the tide of settlement to the Western valleys and mountains.

In dwelling upon this thought I am reminded of a friend who inherited a few thousand dollars from his father, went West, and became a successful lumber merchant. When I last met him he was congratulating himself upon how much he had done with his small patrimony, because he had come West while it was yet in its day of small things. "I thank God I was born in New

INTRODUCTION

England," he said in a fine burst of pride, and then, with a smile of satisfaction, he added, "and I thank God I had sense enough to get up and leave it!" In this connection, it is perhaps worth while to remark that New England is no longer a mere geographical term. It is a certain spirit of civic pride and individual enterprise. In this sense, New England is very much in evidence in the Far West. Its sons go forth to conquer the waste places. They plant their traditions, and raise a crop of institutions.

A word should be said to those who fear that their children will not have educational advantages in the West. The fear is quite groundless, for every Western State and Territory is lavish in its expenditure for education. Their schools are magnificently endowed with the Nation's gift of public land, and the popular pride in their development is boundless. Even the drawbacks of attending school in country districts are, when rightly viewed, seen to be more advantageous than otherwise. They make the fibre of strong men and women, for they drive the children into the open air, generally on horseback. If you are inclined to be pessimistic about the future of the Republic, just watch a cavalcade of Western boys and girls as they gallop to school through sunshine or rain, and you will find your faith renewed. These strong-limbed, sun-browned, self-reliant youngsters are the citizens of the future. You need not fear that they will fail us in time of need.

But it is not only in the common schools that the West is strong. It is strong in its colleges and universities, and stronger yet in the men who stand at the head

INTRODUCTION

of them. They are the men of the Forward Look. They are clearing the intellectual forests, rooting up the social sage-brush, irrigating the arid wastes of politics and economics. Ah, what a harvest they are preparing for the future—the David Starr Jordans, the Benjamin Ide Wheelers, the George A. Gateses, and the rest of the big-brained, big-hearted brood who are training the rising generation in the bright sunshine of the Pacific Coast!

A new era is dawning on the Western half of the continent. The rough edges of pioneer life have worn off, and speculation is giving place to sober industry. The national irrigation policy lends an element of certainty, of stability, which was sadly lacking in the past. When Uncle Sam puts his hand to a task, we know it will be done. Not even the hysteria of hard times can frighten him away from the work. When he waves his hand toward the desert and says, "*Let there be water!*" we know that the stream will obey his command. We know more than that—know when the water will come, how much land will be reclaimed, how many homes will be builded. We can even calculate with precision how many towns will spring up and where they will be, and the railroad actuary can figure out the traffic of the future.

There never was such a time as now for the young man to go West and grow up with the country. It is no longer a wild adventure, but the sane planning of a career. The private capitalist, even after we have wooed and won him, may change his mind, or lose his fortune, or death may arrest him and wreck a thousand hopes. But the public capitalist is dependable. He

INTRODUCTION

does not change his mind, or lose his fortune, and he lives on with the generations. This capitalist has been enlisted—the United States of America, Unlimited!

Go West, young man! That is, if you are the *right* young man, with the Western temperament, and—if your wife is willing!

INTRODUCTION TO THE 1969 EDITION

IT seems especially important to reissue William Ellsworth Smythe's *The Conquest of Arid America* now, when the nation's reclamation era is undergoing a transformation with the phasing out of the "high dam" construction program, and when conservation is attracting a greater following than ever. This book offers its readers a statement of the history, rationale, and program of the movement that launched the reclamation era almost seven decades ago when the Newlands Act was passed (1902). *The Conquest*, first published by Harpers in 1900, became so popular with the inauguration of various Reclamation Service projects that a new and revised edition was undertaken by Smythe and published by Macmillan in 1905, and new printings followed at least to 1911.

This classic of the Progressive Era reclamation movement must be understood in its own optimistic terms as illustrating values in the American heritage to which its author was appealing, as well as the widely recognized social ills that Smythe and like-minded citizens were attempting to reform. The work has still another dimension for those interested in the history of conservation. William Ellsworth Smythe helped mobilize public opinion in support of the reclamation movement as founder

INTRODUCTION TO THE 1969 EDITION

of the National Irrigation Congress and the influential journal, *Irrigation Age,* beginning in 1891. His voice, manifest through articles in national periodicals after he left *Irrigation Age* in 1895, was an influential force for enactment of the 1902 law. He continued to render support to the reclamation movement long after the fledgling Reclamation Service began to implement the dream he shared with John Wesley Powell.

William Ellsworth Smythe ranks as one of the foremost leaders of the federal reclamation movement. He has, however, never received the attention from historians that his vital role in the formative years deserved. And, as the following biographical sketch demonstrates, reclamation was but one phase of Smythe's eventful career as a Progressive Era reformer.

Smythe was born December 24, 1861, at Worcester, Massachusetts, into a family tracing its ancestry back to Governor Edward Winslow of Plymouth Plantation. High school newspaper experience combined with the influence of Parton's *Life of Horace Greeley* interested him in journalism.[1] His father, a wealthy shoe manufacturer, reluctantly permitted the boy to leave home at the age of sixteen, and after serving his apprenticeship with several Massachusetts papers, he assumed the position of editor of the Medford *Mercury* at the age of nineteen.

---

[1] Harriet B. Smythe, "Biographical Sketch," *Irrigation Age,* XIV (October, 1899), 3-5; Bensel R. Smythe, "A Long Memory of Life with Father" (MS, 1968); "William Ellsworth Smythe," *The National Cyclopedia of American Biography* (New York, 1927), XVII, 443-44; George W. James, *Heroes of California* (Boston, 1910), pp. 465-81.

INTRODUCTION TO THE 1969 EDITION

Subsequently he was on the Boston *Herald* staff before he undertook an ill-fated book-publishing venture that led to financial reverses. Perhaps it was Greeley's influence that persuaded the twenty-seven-year-old Smythe to take up the offer of a Nebraska townsite developer and become editor and publisher of the Kearney *Expositor* in 1888.

His autobiographical sketch in the 1905 edition of this book (the version reissued here) discloses the conjuncture of events—the extended drought, an editorial assignment with the Omaha *Bee*, and his visit to New Mexico—which led William Smythe to take up his quest for reclamation.[2] Other influences certainly included the example of the irrigation colony Greeley sponsored in Colorado, as well as Smythe's sense of New Engand town planting history, and the campaign for irrigation in drought-ridden Nebraska in 1890 which had become a national crusade by 1891.

"The cross of the new crusade" was borne by Smythe for the rest of his life. Irrigation came to represent far more than "ditches and acres"; it came, in fact, to be the instrument for transforming society in the western third of the continent. Social reform became his life's passion, and journalism was to be the medium by which

[2] Smythe's *The Conquest of Arid America* is used as the principal source for the genesis of the reclamation movement. George W. James, *Reclaiming the Arid West: The Story of the United States Reclamation Service* (New York, 1917), pp. xv-xvii, 20; Walter P. Webb, *The Great Plains* (Boston, 1931), pp. 356-66; Wallace Stegner, *Beyond the Hundredth Meridian* (Boston, 1954), p. 415; Roy E. Huffman, *Irrigation Development and Public Water Policy* (New York, 1953), pp. 21-26.

INTRODUCTION TO THE 1969 EDITION

he would, like his mentor Horace Greeley, "rouse the Nation to a realizing sense of its duty and opportunity." But he would add new techniques for a new age. As a young man in New England he had demonstrated an ability to organize popular support and to charm audiences with his eloquence and sense of moral earnestness. The convention, the local clubs of concerned citizens, and the broadly based speaking campaigns would be the hallmark of Smythe's future endeavors. More immediately, it was his genius that put together the National Irrigation Congress, which met for the first time at Salt Lake City in 1891. Smythe headed up its organization, shaped its program, and clearly provided its leadership until 1895.[3] Disputes that surfaced from time to time among the leaders of the movement were composed by Smythe's brand of personal diplomacy.[4] The campaign for national reclamation found its most eloquent voice in his journal, *The Irrigation Age,* published first at Salt Lake City and then at Chicago. It was a major force in creating a western public opinion favoring irrigation development. Two years later he expanded his sights to include the national arena. His article in the *Review of Reviews,* "The Irrigation Idea and Its Coming Congress," was the first account of irrigation to be published

[3]Martin E. Carlson, "William E. Smythe: Irrigation Crusader," *Journal of the West,* VII (January, 1968), 41-47; Stanley R. Davison, "The Leadership of the Reclamation Movement, 1875-1902" (Ph.D. dissertation, University of California, Berkeley, 1951), pp. 142-235.

[4]William C. Darrah, *Powell of the Colorado* (Princeton, N.J., 1951), pp. 308-12; Paul W. Gates, *History of Public Land Law Development* (Washington, D.C., 1968), pp. 468-69.

INTRODUCTION TO THE 1969 EDITION

in the periodical press of national circulation.[5]

Colony settlement of arid America was the theme that ran through William Ellsworth Smythe's numerous writings and his speeches before the several irrigation congresses. Although he also advocated state water law reform, the proposal for cession of irrigable lands to the states, and the encouragement of irrigation investment opportunities; Smythe wanted to participate personally in actual colonization in the West. When *The Irrigation Age* began to run into financial difficulties in 1895, Smythe decided then was the time to inaugurate his colony idea. His pioneer venture, the New Plymouth colony in Idaho, is described in *The Conquest* in somewhat more glowing terms than it deserved. Smythe was also associated with northern California colony experiments in Tehama County and at Standish in Lassen County before he removed to San Diego in 1900. Undaunted by his inability to attract settlers and developmental capital in this period following the boom, Smythe felt prompted to address a national audience. The first edition of *The Conquest of Arid America* was written at his desert home in Lassen County. His somewhat imaginative scheme for raising cooperative capital was outlined there.

Smythe's preachings created a particular image of western America in the minds of an eastern audience that was especially receptive at this time to such an image.

---

[5]*Review of Reviews,* VIII (October, 1893), 394-406. There is no bibliography of Smythe's writings, for instance, that would disclose his numerous contributions to *Century, Forum, North American Review, Atlantic Monthly, Review of Reviews,* and, on the Pacific Coast, *Land of Sunshine* and *Out West.*

INTRODUCTION TO THE 1969 EDITION

American society had been traditionally integrated around a system of values growing out of its simple agrarian past, a Protestant religious faith, and an individualistic approach to government and the economy. With the onset of the Panic of 1893 this system of values —long subjected to the deteriorating forces of industrialism, monopoly capitalism, urbanism, and the social tensions arising from the new immigration—suffered a particularly strong shock when the depression deepened and radical political movements challenged the status quo.

The East had traditionally held a romanticized but suspicious attitude toward the West. It was thought of as wild and woolly, the dwelling place of savages, an uninhabitable wilderness and desert. The Populist excitement also heightened fears of the West. By mid-decade, however, and certainly following the 1896 election, interest in the West increased. The image of the West was etched in more positive terms. People like Frederic Remington, Theodore Roosevelt, and Owen Wister became the interpreters of a more friendly West, and a genre of popular writers featuring Mary Austin, Stewart Edward White, and Harold Bell Wright peopled the West with recognizable heroes. This was a West that Charles Lummis and George Wharton James had made familiar to a generation of tourists.[6] At this high tide

[6]G. Edward White, *The Eastern Establishment and the Western Experience* (New Haven, Conn., 1968), pp. 171-202; Franklin Walker, *A Literary History of Southern California* (Berkeley, Calif., 1950), pp. 222-26; Edwin Bingham, *Charles F. Lummis: Editor of the South-West* (San Marino, Calif., 1955), pp. 145-51.

INTRODUCTION TO THE 1969 EDITION

of national interest in the West, William Ellsworth Smythe began to catch the public ear with his message that arid America was hospitable, not alien territory. The overriding theme of *The Conquest of Arid America* was that the West could offer a sanctuary for the traditional American society nourished in agrarian simplicity and protected from the forces that were undermining the old order in the East.

Some recent historians have stressed the role of Senator Francis G. Newlands, the eastern scientific community, and President Theodore Roosevelt in the passage of the Newlands Reclamation Act of 1902.[7] This is a reinterpretation of the long-accepted view expressed within Smythe's book that a western-based movement and a bipartisan band of western congressmen produced the necessary votes to achieve this legislative triumph. Certainly it is in keeping with contemporary views to suggest that East and West agreed on this piece of innovative legislation because of their allegiance to a common symbol, that of the American homestead farmer. Apprehensions over the closing of the frontier, the agitation over public land frauds, and President Theodore Roosevelt's impassioned advocacy of the Newlands Act as a measure that would create more homesteads contributed to a debate in both houses of Congress which rang all the changes on the homestead ideal, a sacrosanct feature

[7]Samuel P. Hays, *Conservation and the Gospel of Efficiency: The Progressive Conservation Movement 1890-1920* (Cambridge, Mass., 1959), pp. 1-15; William Lilley III and Lewis L. Gould, "The Western Irrigation Movement, 1878-1902: A Reappraisal," *The American West: A Reorientation,* ed. Gene M. Gressley (Laramie, Wyo., 1966), pp. 57-74.

INTRODUCTION TO THE 1969 EDITION

of the American covenant.[8]

The first edition of *The Conquest of Arid America* explained how this traditional homestead ideal could be made especially applicable to the desert regions of arid America. "Manless land" in the West cried out for the "landless man" made surplus in eastern cities. Arid lands were phenomenally fertile and could be cultivated intensively in small tracts "under the irrigation ditch." The homestead ideal could be fully realized in the West, where colonies of homesteaders would shed the speculator from their midst, live in a state of democratic equality with culture-enhancing institutions close at hand. In this context the homestead symbol was potent enough to convince the conservation-shy westerners that reservation of public domain land, when carried out on a small scale around future reservoir sites, was an acceptable plan.

Smythe's brief homesteading experience in the sagebrush desert country of northeastern California had a marked effect upon his future career and the reclamation cause. Well before the adoption of the Newlands Act, he learned the practical difficulties faced by a colonization project such as his town site of Standish in the Honey Lake region. Private investment failed; there was no survey of water resources; state water laws were conflicting; only the federal government could afford the expense of reservoirs. These points were amply developed in Smythe's subsequent California campaign to reclaim the deserts of the West. Fortune finally smiled upon his

[8] E. Louise Peffer, *The Closing of the Public Domain: Disposal and Reservation Policies, 1900-50* (Stanford, Calif., 1951), pp. 38-41.

INTRODUCTION TO THE 1969 EDITION

endeavors. Long-time acquaintanceship with the national irrigation engineering fraternity brought Smythe a staff appointment to Elwood Mead's Irrigation Institutions survey of 1900. His study of Honey Lake Basin incorporated recommendations for a new California water code.[9] That same year Smythe established a new base of operations when he was appointed to an editorial staff position on the San Diego *Union*. Here he joined forces with the California Water and Forest Association and, as vice-president of that organization, stumped the state for a reform of the state water code modeled after the Wyoming statute, with determination of water rights reposing in the state government.[10] The adoption of the federal reclamation law offered a powerful argument for public construction of all reservoirs in the state, but the Smythe campaign for codification ran into strong opposition from the private water monopoly interests.

The national reclamation cause received additional support after Charles Lummis invited Smythe to run a department, "The Twentieth Century West," in the influential West Coast magazine, *Land of Sunshine* (later, *Out West*).[11] Smythe continued to mark the progress of federal reclamation, to bemoan the small sums that were available under the Reclamation Fund,

[9] William E. Smythe, "The Irrigation Problems of Honey Lake Basin, California," *Report of Irrigation Investigations in California*, directed by Elwood Mead (Washington, D.C., 1902), pp. 71-113.

[10] *Water and Forest: An Illustrated Magazine*, I (November, 1900), 10; *Land of Sunshine*, XVI (January, 1902), 78-79.

[11] Charles Lummis, "The Lion's Den," *Land of Sunshine*, XIV (June, 1901), 511-12.

INTRODUCTION TO THE 1969 EDITION

to support the decision of the Secretary of the Interior to provide Reclamation Service water to private landholders, and to advocate new formulas to increase the number of projects in the West.[12] He celebrated the completion of the first federal dam and, at considerable labor and with the fullest cooperation of the Department of Interior, presented for publication a completely revised edition of *The Conquest of Arid America*. The last two chapters of this book reflected the esteem he held for the work of "Uncle Sam's young Men" in the Reclamation Service. It probably never occurred to him that these projects he lauded were to cost the nation more than it would receive in agricultural benefits.[13] As self-appointed publicist for national reclamation, Smythe was committed to fashioning a great propaganda campaign on behalf of a restored homestead ideal in arid America under federal government auspices.

Smythe played other roles in the Progressive Era. His travels throughout the state on behalf of the California Water and Forest Association had brought him into contact with groups of concerned citizens whom he now organized into clubs federated into the California Constructive League. He wished to challenge both parties with a program for "Twentieth Century Constructive

[12]Bingham, *Charles F. Lummis,* p. 70; William E. Smythe, "A Success of Two Centuries," *Out West,* XXII (January, 1905), 72-76.

[13]Vernon W. Ruttan, *The Economic Demand for Irrigated Acreage: New Methodology and Some Preliminary Projections, 1954-1980* (Baltimore, 1965); Otto Eckstein, *Water-Resource Development: The Economics of Project Evaluation* (Cambridge, Mass., 1965), pp. 226-36.

INTRODUCTION TO THE 1969 EDITION

Democracy.'' Almost by accident William Smythe found himself Democratic candidate for congressman from the San Diego district in 1902. Though he lost the race he won the lasting friendship of the Democratic candidate for governor, Franklin K. Lane. The campaign proved to Smythe that there were special interests other than water monopoly that he must oppose. During the course of his campaign and subsequently on lecture platforms and in the pages of *Out West,* he continued to focus attention on the monopoly problem with its attendant antisocial business practices, concentration of wealth, and corruption of government. The solutions offered included producer cooperatives, compulsory arbitration of labor disputes, and the leasing of large estates.[14]

For a time Theodore Roosevelt's Square Deal was the inspiration for Smythe's campaign against special interests in California. Smythe succeeded in bringing his many suggestions for social reform together within the covers of a book, *Constructive Democracy: The Economics of a Square Deal.*[15] He argued for a system of federally chartered corporations and for governmental regulation of business profits. This program embodied the current views of Senator Newlands and of experts in the Roosevelt administration, such as James Garfield of the Bureau of Corporations, who encouraged consolidating American industries in the interest of efficiency. Smythe here reflected the visionary reformer who

[14] William E. Smythe, "Running for Congress," *Out West,* XVII (December, 1902), 758-65.
[15] William E. Smythe, *Constructive Democracy: The Economics of a Square Deal* (New York, 1905).

INTRODUCTION TO THE 1969 EDITION

relied on the federal government to curb injustice. In asking government to control all national industry and set limits to the profits of such enterprise, Smythe failed to appreciate the dynamics of the capitalistic system and the then acceptable limits of governmental intervention.

It was to be expected that this California reformer would find the Progressive state administration of Hiram Johnson congenial to his interests. In 1912, however, in spite of his arguments in *Constructive Democracy*, he returned to his former party and campaigned actively for Woodrow Wilson. Theodore Roosevelt seemed too closely associated with the steel trust. The shift in allegiance was symbolic of the true tenor of Smythe's commitment to a simple agrarian society of yesteryear. Smythe's agrarian democracy had already attracted him to the "back-to-the-land movement." His California-based Little Landers colonies, starting with San Ysidro near San Diego and then expanding to other colonies at Tujunga, Runnymede, and Hayward, revealed Smythe as founder, resident promoter, and author who advocated a new way of life in association with the "New Earth."[16] Once again a cause, this time one that was divorced from the fostering hand of government, led to the production of a journal and a book. The magazine, entitled *Little Lands in America* and edited by his wife, Harriet Bridge Smythe, began publication in 1916. It conveyed the message that latter-day home-

[16]Henry S. Anderson, "The Little Landers' Land Colonies: A Unique Agricultural Experiment in California," *Agricultural History*, V (October, 1931), 139-50.

INTRODUCTION TO THE 1969 EDITION

steaders could link together the virtues of urban and rural living by dwelling on one-acre subsistence lots grouped into hamlets and drawing upon the cultural riches of metropolitan areas. Smythe's enthusiasm never waned, though the colonies failed to achieve the success he had envisioned for them.

This chapter in Smythe's life came to an end following the death of his wife when he accepted a position in Washington under Interior Secretary Lane.[17] Smythe assisted Lane in the much discussed but abortive soldiers' homestead colony program and then launched a new colonizing program under the sponsorship of his American Homesteaders Society to encourage working men to purchase homes of their own in suburban tracts. His last book, *City Homes on Country Lanes* (1921), incorporated much of the practical experience Smythe had gained as colony promoter.[18] On October 5, 1922, Smythe died, probably without knowing that the back-to-the-land movement had lost its appeal for most of his fellow Americans.

The Conquest of Arid America was Smythe's most memorable book. One can find within its pages most of the values that motivated his subsequent career. It is easy to picture him as a latter-day New England Puritan whose conscience insistently urged him to battle against

[17] Bensel R. Smythe, "A Long Memory of Life with Father"; Franklin K. Lane, "The Returning Soldier," *Reclamation Record*, X (January, 1919), 3-8; Bill G. Reid, "Franklin K. Lane's Idea for Veterans' Colonization, 1918-1921," *Pacific Historical Review*, XXXIII (November, 1964), 447-61.

[18] William E. Smythe, *City Homes on Country Lanes: Philosophy and Practice of the Home-in-a-Garden* (New York, 1921).

INTRODUCTION TO THE 1969 EDITION

social injustice. His appreciation for Mormon institutions and deliberate effort to give Brigham Young and the Mormons a good press when it was hardly fashionable so to do can be explained in part by his affinity for Yankee-type Mormon villages. His real metier was as promoter of advanced reforms in the interests of social justice. Thus he popularized European cooperatives, New Zealand instruments of social democracy, and Zionist desert colonies in Palestine, as well as national reclamation and Little Landers colonies. As a social reformer, he shared the common Progressive Era values. He believed in a universal world order governed by laws, a benevolent Deity, and the duty of man to work in partnership with God in perfecting this world.[19] In the West, Smythe saw the desert as the crucial factor in bringing into being that well-ordered society.

William Ellsworth Smythe was a newspaper and periodical publisher, editor and feature writer, author of an excellent *History of San Diego,* a real-estate promoter, a politician, a public servant, a friend of governors and presidents, and a friend to mankind.[20] His fast-paced career, while scarcely mirroring a procession of triumphs, did bring him an ever widening circle of friends. Friendships, not material rewards, were his recompense for a life of dedication. The towering monu-

[19]Clarke A. Chambers, "The Belief in Progress in 20th Century America," *Journal of the History of Ideas,* XIX (April, 1958), 197-224.

[20]William E. Smythe, *History of San Diego, 1542-1907: An Account of the Rise and Progress of the Pioneer Settlement on the Pacific Coast of the United States* (San Diego, 1907); San Diego *Evening Tribune,* April 27, 1959, p. B-1.

INTRODUCTION TO THE 1969 EDITION

ments in concrete built as Reclamation project dams in our West reflect our nation's debt to Smythe. If his championship of a reincarnated homestead ideal seems dated today, one must recognize the strength of the agrarian heritage in Progressive Era America and Smythe's earnest desire to preserve the best of America's heritage in a world of fast-moving change.[21]

LAWRENCE B. LEE

San Jose, California
January, 1969

---

[21] Paul K. Conkin, *Tomorrow a New World: The New Deal Community Program* (Ithaca, N.Y., 1959), pp. 18-23; Wayne C. Rohrer and Louis H. Douglas, *The Agrarian Transition in America* (Indianapolis, 1969), pp. 25-47.

THE CONQUEST OF ARID AMERICA

Part First

COLONIAL EXPANSION AT HOME

"In 1850 she [the United States] passed Austria. In 1860 it was her motherland to whom she held out her hand lovingly as she swept by. In 1870 she overtook and passed France. In 1880 she had outstripped the German Empire ; and now, in 1890, she is left without a competitor to contend with except giant Russia. All the others she has left behind. Another decade, and the sound of the rushing Republic close behind will astonish even Russia, with its eighty-six millions in Europe. Yet another decade, and it, too, like all the rest, will fall behind to watch for a time the new nation in advance, until it forges so far forward as to pass beyond her ken, when five hundred millions, every one an American, and all boasting a common citizenship, will dominate the world—for the world's good."—ANDREW CARNEGIE, *Triumphant Democracy.*

CHAPTER I

GREATNESS BY CONTINENTAL CONQUEST

THE economic greatness of the United States is the fruit of a policy of peaceful conquest over the resources of a virgin continent. Without this great item of raw material, the finished product which the world acknowledges in the industrial America of to-day would have been impossible.

The true career of the American people as a race of empire-builders dates not from the founding of Jamestown, New Amsterdam, and Plymouth, but from the surrender of Cornwallis at Yorktown and the subsequent inauguration of George Washington as the first President of the United States. The early settlers were merely European sentinels standing guard over a treasure of continental magnitude which they neither comprehended nor appreciated. The tobacco-raisers of Virginia, the fur-traders of New York, and the religious enthusiasts of New England had no conception of a national destiny or mission. They looked backward to the civilization whence they had come, rather than forward to the conquest and subjugation of the mightier empire on whose eastern shores they had set their reluctant feet.

Only at the close of the successful war for indepen-

THE CONQUEST OF ARID AMERICA

dence did the world begin to realize that the American was to be the master of the new continent for all time, and that his rule must move westward as naturally and inevitably as the sun in its course. Only when the new government, hewn out with the sword and cemented with the blood of its citizens, had been finally and firmly established, did the heterogeneous elements in the sparsely settled original States crystallize into a national spirit and a national character. From that hour the material development of the New World began in earnest. The people labored as with the vim and courage of him who works for himself. Men began to dream of an America which should be richer and more populous and powerful than even Europe.

The war was over—the war was begun! England had been shaken off by force of arms, and the other European ties would be loosed by the arts of diplomacy; but it remained to wage war on the forest, the plain, the desert, and the mountain, and to create a better civilization than the world had seen. What millions of men and billions of dollars were employed and rewarded in the process—what workshops, and railroads, and farming districts were created in the wilderness—what cities, with swarming thousands of inhabitants, with homes and colleges and hospitals, were erected in the midst of the primeval silence—what States were carved from the woods and prairies—what unexpected commerce, borne in undreamed-of steamships, was sent to whiten the unexplored inland seas!

It is in the answer to these questions rather than in the poet's pæan to democracy that the true explanation of the economic progress of the nation will be found.

CONTINENTAL CONQUEST

It is not to be denied that the fact that the United States was heralded throughout the world as a "free country" attracted millions of immigrants, nor that popular government and complete immunity from the demands of royal tribute left enterprise unhampered to a degree hitherto unknown. But a vast commerce can no more find sustenance solely in the written constitution of a country than a starving prospector in the mountains can satisfy his appetite with scenery.

It seems worth while to lay strong emphasis upon this point, because the somewhat general acceptance of the notion that America is the product of its institutions, rather than that its institutions are the product of America, has obscured the causes of past prosperity and belittled the importance of our undeveloped resources. Not until this fact is understood and acknowledged is it possible to comprehend, even vaguely, the incalculable importance of the undeveloped regions in the western half of the United States.

At the close of the Revolution the United States consisted of a fringe of settlements mostly confined to the Atlantic coast and the banks of important rivers on the eastern slope of the Alleghanies. Nominally, the national domain extended westward to the Mississippi river, but practically there was no development beyond the thirteen original States. Even there the natural resources of the country had scarcely been touched. Boston had a population of about eighteen thousand, New York of about thirty thousand, Philadelphia of about thirty-five thousand, Baltimore of about fifteen thousand. Richmond, Charleston, and Savannah, though of some importance politically, were mere straggling hamlets.

THE CONQUEST OF ARID AMERICA

Detroit, St. Louis, and New Orleans were French outposts in the wilderness. Buffalo, Cincinnati, Cleveland, and Chicago; Omaha, Kansas City, Denver, Salt Lake, and San Francisco—these and scores of other cities now populous and powerful — were hidden in the womb of time. Of the country between the Alleghanies and the Mississippi far less was known than the world now knows of Africa. The vast domain lying between the Father of Waters and the Pacific Ocean was neither as well explored nor as perfectly comprehended as the Arctic region is to-day.

When the men of the new Republic turned their backs on the Old World, in the double sense of politics and industry, and faced the continental opportunity which awaited them, they entered upon the fiercest war of conquest in all history. And the spoils of that war were to be in proportion to the magnificence of the task.

The first effort at the subjugation of the wilderness was directed to the fields and the streams. The forest clearings were extended that agriculture might find room for expansion. The trees felled in the process were floated in the rivers to saw-mills driven by the current. The logs, transformed to lumber, supplied the material for millions of comfortable homes. In the mean time, the new farms fed the growing population of the towns, while a myriad of workshops, improved by inventions of which a robust necessity was the prolific mother, consumed and manufactured the textile materials from field and pasture.

The step from the crude employments of the frontier to the manifold occupations of a modern industrial life was easy and natural. Fostered by a generous policy of protection, and blessed by long years of peace, the work of development went on from generation to generation. In New England the raw material on which the workmen labored in fashioning a civilization was poorer than elsewhere. And yet it was on that sterile soil, in the midst of those rocks and hills, that industrial pre-eminence was first to be achieved. A citizen of Massachusetts once made the just boast that "not one drop of water flows from our hills to the sea until its power has been three times multiplied by the mill wheels." Every stream was lined with factories, nearly every town had its peculiar industries and its growing crowds of skilled laborers, supporting the stores and shops with their trade, and filling the schools with their children.

Not only in New England, which owed its serious energy to the example and character of its founders, and its fierce industrial enthusiasm to a system of free labor, but equally in New York, in New Jersey, in Delaware, in Pennsylvania, and southward to the Floridian peninsula, the army of labor marched on with irresistible advance. It scaled the crests of the Alleghanies and opened yet greater valleys to the energy of men. It tunnelled into the earth and brought up the hidden stores of coal and iron ore. It tapped the subterranean reservoirs of natural gas and oil.

With the rapid growth of a many-sided economic life the need of improved facilities for internal transportation arose and grew yearly more urgent. The natural watercourses, navigated by rafts and sailing craft, did not long suffice. The army of labor was set at work in building great highways and digging canals. Then came the steamboat, and, finally, the railroad with its iron horse.

THE CONQUEST OF ARID AMERICA

Thus it was that the work of taming the wilderness went on with increasing fervor. Thus it was that thirty-two new States were added to the original thirteen. Thus it was that the national population was increased fourteen-fold, and that cities rivalling the greatest urban centres in the Old World, in size and wealth and power, were developed on the site of the colonial villages of the early days. Thus it was that the Republic was able to welcome, and to absorb into its apparently insatiable industrial system, the millions of immigrants who flocked to its shores.

During these days of rapid material expansion over new areas, Uncle Sam was the proprietor of the most gigantic employment bureau on earth. He had enough work for his own prodigious family of sons, and for the overflow of all the families across the sea. He offered the highest wages in the world-wide market. He distributed his abounding prosperity through all channels of trade, all classes of industry, all grades of society. He made men and communities rich first by employing their energies in the conversion of the wilderness into a civilization, and paying them roundly for the work; then by the rise in values, or "unearned increment," which comes with population and development; finally, by the premium, or interest, upon capital thus acquired. All this was the logical fruit of a policy of continental conquest bravely undertaken, magnificently achieved.

Behold the story of national prosperity in the form of a few clear-cut figures, divested of all rhetorical clothing: In a little more than one hundred years the area of farms increased from sixty-five thousand square miles to over one million square miles. The number of persons engaged in the agricultural industry in 1900 was 10,438,-219,—more than two and one-half times the entire population in 1790. In acres the total amount of land classed as farms by the last census was 841,201,546, of which 414,793,191 acres were under actual cultivation, the rest being woodlands. The number of individual farms was 5,739,657. The annual product was worth $3,764,177,706. "In ten years," says Mr. Andrew Carnegie, in his inspiring book, *Triumphant Democracy*, "a territory larger than Britain, and almost equal in extent to the entire area of France and Germany, was added to the farm area of America."

Marvellous as this statement is, it exhibits but one item in the record of continental conquest which conferred such phenomenal prosperity upon the American people in the past. Agriculture is the basis of civilization, and upon the foundation so quickly and thoroughly laid, the new nation hastened to erect the superstructure of a complex industrial life. The existence of an enormous population on the farms furnished a great field for manufactures. This industry now employs between five and six million workmen, who annually receive and expend over two billion dollars in wages and create an annual product worth thirteen billion dollars.

Agriculture and manufactures—both finished products wrought by millions of workmen from the raw materials of the new continent—combined in demanding the most extensive arrangements for internal transportation ever provided on the face of the earth. The total railroad mileage at the last census was one hundred and ninety-four thousand three hundred miles, which is more

THE CONQUEST OF ARID AMERICA

than that of all European countries combined. Of this vast mileage, forty-eight per cent. was built before 1880, thirty-eight per cent. between 1880 and 1890, and fourteen per cent. between 1890 and 1900. When it is remembered that each of these miles stands for about fifty thousand dollars expenditure—the cost of construction and equipment—and that the work employed an army of laborers and skilled artisans, who in turn consumed great quantities of agricultural and manufactured products, it is not difficult to realize that the railroad development contributed largely to the national prosperity in the past. It was, of course, the direct result of the great process of material conquest which was going on.

To the same cause was due the employment of nearly five million people in trade and transportation; of a million and a quarter in professional services; and of nearly four hundred thousand in mining. The grand result is seen in the fact that the national population grew from less than four millions in 1790 to more than seventy-six millions in 1900, while the total wealth mounted to the incomprehensible sum of ninety-four billion dollars.

Such are the stupendous results of the labors of a great people applied to the resources of a virgin continent. Other people have possessed energy and genius, and two of the European nations have enjoyed the blessings of self-government. If republican institutions would alone guarantee such results in the future, it is hardly to be imagined that the sternest monarchy could withstand the demand for their adoption. But the transcendent

CONTINENTAL CONQUEST

factor in the result was the continental expanse of marvellous resources awaiting the labor and genius of man.

Can there be any question that the abounding prosperity of the American people during the first century of their national life was due to this luminous fact? Can there be any reasonable doubt that if the policy of material conquest over new areas can find another field on which to operate, and that if it be entered upon with the old vigor and faith, it will confer another century of prosperity upon the nation so fortunately endowed?

CHAPTER II

THE HOME-BUILDING INSTINCT OF THE AMERICAN PEOPLE

SPEAKING in broad terms, there have been three great eras of colonization in the United States. All of these eras have been well defined, intelligible, and eventful. They peopled successively the Atlantic coast, the trans-Alleghany region from Lakes to Gulf, and the valley of the Mississippi. Taken together, they made virtually complete the conquest of Eastern America, and in Eastern America over ninety per cent. of the national population dwells to-day.

A study of these historic movements reveals a striking fact. It is a fact which throws a flood of light on the American character, explaining much that has occurred in the past and furnishing secure ground upon which to base predictions of much that is to happen in the future. The American colonist, from Plymouth in Massachusetts to Plymouth in Idaho, has fixed his eyes on one star, which has shone out serene and steady through the clouds of religious persecution, of war, and of economic strife. That star stood for home. To build a home for himself and his children, to live there at peace with his neighbors and the world, to make better institutions for average humanity—this, when the subject is viewed as a whole, is seen to have been the con-

A FINE EXAMPLE OF FLOWING ARTESIAN WELLS; HEAD OF RIVERSIDE CANAL SYSTEM, CALIFORNIA.

THE HOME-BUILDING INSTINCT

sistent aim of American colonization from the beginning. There are a few exceptions to be noted, but they are not of sufficient importance to affect the general result. Such exceptions are the settlement of California, and of certain localities in the Rocky Mountains, during periods of excitement following the discovery of gold. Another instance was the settlement of Kansas as a means of preserving the equilibrium between the free and the slave States. But these are isolated instances, of far more moment in an historical than in a numerical sense. The settlers of the United States have been moved by very different instincts and motives than those which impelled the Romans, the Normans, and Danes to settle at different periods in Britain. The great movements of population in the Middle Ages were armed conquests for spoils, and power, and martial glory. Those, indeed, were the ruling motives among Europeans and Asiatics until comparatively recent times. When these motives ceased to operate, they were succeeded by another which was equally sordid, even if more humane. This was the lust for trade or for sudden riches. This it was which impelled the settlement of Australasia by the English, of the Spice Islands by the Dutch, of South America by the Portuguese, of Cuba by the Spanish, of Africa by all of these and by the French and Germans as well. Thus the hosts which swarmed out of Europe to make new settlements all over the earth were principally marshalled under the flag of avarice. It was far different with the men who, at various periods during the last three hundred years, conquered the soil of the United States and extended the frontiers of its civilization.

THE CONQUEST OF ARID AMERICA

The settlement of the New World was largely inaugurated by those who fled from religious persecution. But it cannot be said on that account that their ruling motive was not the desire to enjoy the security of a home. Religious sentiment lies very close to the hearth-stone. Upon its human side, at least, it has nothing in common with politics. Still less is it related to the struggle for gain. It was because they could not live at peace in Europe, because they could not be certain of life or tenancy in any one place, and therefore could not accumulate a competence for their children, that the religious enthusiasts fled over the sea. The Puritan in Massachusetts, the Baptist in Rhode Island, the Quaker in Pennsylvania, and the Catholic in Maryland, looked less passionately upon their spires and crosses than upon the babies in their cradles, the vegetables in their gardens, and the smoke which curled from their chimneys.

It is true that there were many fanatics in the seventeenth and previous centuries to whom religion was dearer than home; but it was not the axes of these fanatics that felled the American forests. Their devoted spirits were freed at the stake, or at the block, or their poor bodies festered in foul prisons. It was the element whose love of home and kindred was too powerful to permit them to suffer martyrdom, even though their convictions forbade them to eschew their religious practices, who inaugurated the first era of colonization on these shores. Theirs are the first footprints in our history, and they lead straight to the home and the fireside.

The second real era of colonization came with the end of the Revolution. Previous to that event the

THE HOME-BUILDING INSTINCT

trans-Alleghany country was but vaguely known as a whole. Daniel Boone had, indeed, built his cabin in the wilds of Kentucky, and adventurous spirits had begun to follow him from Virginia and the Carolinas. James Robertson and John Sevier, leading the hardy backwoodsmen of the Scotch Presbyterian faith, had begun the making of Tennessee. The French Creoles had lived for three generations in the slumberous repose of widely scattered villages in the Ohio Valley, and had gathered in some numbers at New Orleans. But the hour for the real movement of population to the westward of the mountains had not struck. When it did strike, it found the home-building instinct of the American people instantly and passionately responsive to its summons. It was the returning veterans from the War of Independence who lent the first great impulse to the new emigration. Hardened by years of out-door life, thoroughly weaned from the atmosphere of the town and the shop, finding their places on the farms largely filled by boys who, during their absence, had grown to self-reliance, if not to manhood, these war-worn veterans were not unwilling to transfer their battle-ground from the sea-coast to the wilderness, and to fight for homes as ardently as they had struggled for political independence.

During the next thirty years the population of Kentucky leaped from about seventy thousand to over half a million, and that of Tennessee from thirty thousand to over four hundred thousand. Ohio, Indiana and Illinois, which had no place in the census of 1790, were credited, respectively, with nearly six hundred thousand, one hundred and forty-seven thousand, and fifty-five thousand, in 1820. The movement went on without pause until the outbreak of the great rebellion. It was even more plainly marked with the home-seeking character than the earlier settlement of the seaboard States. We need not in this instance seek the home-loving instinct under the religious motive. The circumstances and the methods of the new army of settlers revealed the supreme object of their emigration.

The lands along the coast and in the rich valleys of tidal rivers had been well occupied by a people who enjoyed substantial prosperity, not only as the reward of their industry, but also as the result of their priority of settlement. The country had grown. It was plainly upon the verge of a larger and more rapid expansion. These circumstances enhanced the value of property and laid the foundation of many family fortunes, especially where the colonial hamlets had grown to be towns, and promised to become populous cities. The early-comers and their descendants were being steadily enriched by the unearned increment. Those who were thus established had no occasion to move, but their less fortunate neighbors longed for homes of their own, and were ready to take quick advantage of the opportunity which the war and the Ordinance of 1787 had opened for them in the West. These people were almost universally poor in a worldly sense, but rich in courage and intelligence and full of the spirit of empire-builders. They were no more a class of greedy speculators than were the pioneers of New England. They emigrated in order that they might improve their condition. They were home-seekers pure and simple. Placed completely beyond the influence of Europe, and acting under a new

THE HOME-BUILDING INSTINCT

spirit of nationality, the people concerned in our second era of colonization developed a rugged Americanism before unknown. This spirit was typified in the character of Abraham Lincoln, who was one of its products.

The third era of colonization followed the War of the Rebellion, as the second had followed the War of the Revolution, and largely for the same reason. The cessation of hostilities and the disbandment of the armies turned back into the paths of peace hundreds of thousands of veterans. They were filled with an over-mastering desire for homes. They longed for a chance to work for themselves, as their fathers and forefathers had done. Uncle Sam was still proprietor of a vast estate of virgin and fertile soil. The homestead law beckoned to the returning hosts like the finger of fate. The result was the phenomenal settlement of the Upper Mississippi Valley and the creation of States where the old soldier reigned all but supreme. In a period of twenty years after the war Nebraska jumped from a population of twenty-eight thousand to nearly half a million; Kansas from one hundred thousand to a round million; Iowa from six hundred thousand to a million and six hundred thousand; Dakota from five thousand to one hundred and forty thousand, while Minnesota also added more than half a million to her total.

The movement never paused until it encountered an obstacle beyond the power of the individual settler to overcome. This obstacle was aridity—the failure of rainfall to meet the demands of agriculture. The impetus of the movement carried its vanguard across the danger-line and into the territory where existence could not be maintained without recourse to methods then little understood, and indeed not fully developed. Upon this strange boundary of prosperity, which nature had marked with indelible lines, the hosts engaged in the third colonization era trembled and hesitated for several years, then fell back baffled and disappointed.

The first act in the drama of American settlement ended in the eastern foothills of the Alleghany mountains about 1770; the second, in the neighborhood of the Mississippi river about 1860; the third, midway on the plains of Dakota, Nebraska, Kansas, and Texas about 1890. For each of these historic periods we might find a fit and speaking emblem in its characteristic means of transportation. The emblem of the first would be the little *Mayflower*, tossing on the billows of the Atlantic; that of the second, the heavily laden packhorse, threading his tortuous way through the tangle of the untrodden forest; that of the third, the prairie schooner, steering for the setting sun across the trackless sea of the plains.

The wonderful drama of American colonization has reserved a fourth and crowning act, for which the scenery is arranged and the actors ready.

CHAPTER III

THE BETTER HALF OF THE UNITED STATES

THE ninety-seventh meridian divides the United States almost exactly into halves. East of that line dwell seventy-five million people. Here are overgrown cities and over-crowded industries. Here is surplus capital, as idle and burdensome as the surplus population. West of that line dwell five or six millions—less than the population of Pennsylvania, and scarcely more than that of Greater New York. And yet the vast territory to the West—so little known, so lightly esteemed, so sparsely peopled—is distinctly the better half of the United States.

The West and East are different sections, not merely in name and geographical location, but in physical endowments and fundamental elements of economic life. Nature wrote upon them, in her own indelible characters, the story of their wide contrasts and the prophecy of their varying civilizations. To the one were given the advantages of earlier development, but for the other were reserved the opportunities of a riper time. It was the destiny of the one to blossom and fruit in an epoch distinguished for the accumulation of wealth, with its vast possibilities of evil and of good. It was the destiny of the other to lie fallow until humanity should feel a nobler impulse; then to nurse, in the shadow of its ever-

THE CONQUEST OF ARID AMERICA

lasting mountains and the warmth of its unfailing sunshine, new dreams of liberty and equality for men.

That this is not the popular conception of the mission of the Far West may be frankly acknowledged. The region is little known to the great middle-classes in American life. It has been demonstrated by actual statistics that only three per cent. of our people travel more than fifty miles from their homes in the course of a year. Those who make extended pleasure tours gravitate not unnaturally to Europe, drawn by the fascination of quaint foreign scenes and the fame of historic places. But the comparatively few whose business or fancy has taken them across the continent fail, as a rule, to grasp the true significance of the wide empire which stretches from the middle of the great plains to the shores of the Western sea.

It is a common human instinct to regard unfamiliar conditions with distrust. The first settlers in Iowa engaged in desperate rivalry for possession of the wooded lands, thinking that no soil was fit for agricultural purposes unless it furnished the pioneer an opportunity to cut down trees and pull up stumps. "Land that won't grow trees won't grow anything," was the maxim of the knowing ones. Their fathers had cleared the forests on the slopes of the Alleghanies to make way for the plough and the field, and the new generation could not conceive that land which bore rich crops of wild grasses and lay plastic and level for the husbandman to begin his labors, could have any value. A great deal of hard work was wasted before it was discovered that nature had provided new and superior conditions in the land beyond the Mississippi.

BETTER HALF OF THE UNITED STATES

So it generally happens that the casual Western traveller, looking at the country from car-windows in the intervals between his daily paper, brings back more contempt than admiration for the economic possibilities of the country. One must live in the Far West to begin to comprehend it. Not only so, but he must come with eager eyes from an older civilization, and he must study the beginnings of industrial and social institutions throughout the region as a whole, to have any adequate appreciation of the real potentialities of that half of the United States which has been reserved for the theatre of twentieth-century developments. To all other observers the new West is a sealed book.

The West is divided from the East by a boundary-line which is not imaginary. It is a plain mark on the face of the earth, and no man made it. It is the place where the region of assured rainfall ends and the arid region begins. There have formerly been some costly doubts about the precise location of this line, but these have been dispelled by experience, and the lesson learned in hardship and impressed by disaster is learned for all time. The momentous boundary-line is that of the ninety-seventh meridian, which cleaves in twain the Dakotas, Nebraska, Kansas, Oklahoma, and Texas. East of this line there is a rainfall which is accepted as reliable, though there are alternate disasters of drought and flood, varying in their effects from short crops to total failures.

Even in humid regions nothing is so uncertain as the time and amount of the rainfall. In the whole range of modern industry nothing is so crude, uncalculating, and unscientific as the childlike dependence on the mood of

THE CONQUEST OF ARID AMERICA

the clouds for the moisture essential to the production of the staple necessities of life.

The distinguishing characteristic of the vast region west of the ninety-seventh meridian is, then, its aridity—the lack of rainfall sufficient to insure the success of agriculture. The new empire includes, in whole or in part, seventeen States and Territories. It is a region of magnificent dimensions. From north to south it measures as far as from Montreal to Mobile. From east to west the distance is greater than from Boston to Omaha. Within these wide boundaries there are great diversities of climate and soil, of altitude and other physical conditions.

The arid region was the latest acquisition of national territory, except Alaska, until the late war with Spain. It was unknown and undisputed as late as the Revolution. It was the fruit of James Monroe's negotiations with Napoleon I., resulting in the Louisiana purchase ; of the forcible conquest from Mexico ; of the annexation of Texas, and of the Gadsden purchase in 1853. Unlike the rich and well-watered lands in the valley and around the mouth of the Mississippi, the acquisition of the arid region was not compelled by the irresistible pressure of the frontiersmen. It came as a perquisite with the purchase of Louisiana, and as a concession to manifest destiny. Between the day of its acquisition by the United States and the dawn of its peculiar and enduring civilization, the country was destined to pass through three distinct eras. The first was that of the hunter and trapper; the second, that of the cowboy and the rude miner ; the third, that of the railroad, the land-boomer, and the speculative farmer, with mining reduced to a stable industry.

BETTER HALF OF THE UNITED STATES

The first exploration of the strange new land of the mysterious West owed its initiative to the public spirit of President Jefferson. He had, indeed, but the vaguest conception of the possible utility of the country, and realized that its development would come long after he should have passed from the stage of events. But he was a patron of science, and felt, moreover, a patriotic curiosity to learn what sort of a property the nation had acquired. Congress cheerfully authorized the expedition which Jefferson proposed. The result was the journey of the famous explorers Lewis and Clark, begun in May, 1804. Starting from St. Louis, they ascended the Missouri river to its sources, crossed the Rocky Mountains in Montana, and followed the Columbia river to its outlet in the Pacific Ocean. When they returned and presented their report, the public obtained its first glimmering of knowledge concerning the geology, climate, and animal and human life of the Far West. The subject was then one of remote interest to the nation, which had scarcely acquired its foothold, through actual settlement, on the northwestern Territories between the Alleghanies and the Mississippi.

The second notable explorations were those of Zebulon Pike, which developed a superficial knowledge of Colorado and Mexico. Then came Bonneville, Frémont, and their contemporaries and successors, with adventurous settlers and hardy gold-hunters treading close upon their heels, and effecting little substantial development for decades. Francis Parkman, fresh from college, roamed through the country as far as the Black Hills and old Fort Laramie in 1847–8, and left a lively account of the savage wilderness in *The Oregon Trail*.

THE CONQUEST OF ARID AMERICA

Thus gradually, and attended by many misrepresentations and strange misconceptions, which inevitably scattered wide the seeds of prejudice, the arid region emerged from absolute obscurity and stood partially revealed to men. It was not, however, until a few pioneer settlements had demonstrated undreamed-of results, nor until Major John W. Powell, by utterances as daring as his explorations, had furnished a scientific basis for a brood of new hopes, that the real character of Arid America began to glow, like the belated sun through a morning fog, upon the popular imagination.

The superiority of the western half-continent over its eastern counterpart may not be expressed in a word. It is, rather, a matter for patient unfolding through a study of natural conditions over wide areas, and a scrutiny of the human institutions which are the inevitable product of this environment. Aridity, in the elementary sense, is purely an affair of climate. That it is also the germ of new industrial and social systems, with far-reaching possibilities in the fields of ethics and politics, will be demonstrated further on in these pages. But the first item of importance in the assets of the new West is climate.

When an inhabitant of the Atlantic seaboard, or of the shores of the Great Lakes, or of the lowlands of the South, can no longer withstand the penetration of cold, damp winds, or the malarious breath of swamps, his family physician sends him to the arid West. Throughout its length and breadth it is one vast sanitarium. Its pure, sweet air and sunny skies are instinct with the breath of life. They put new heart into the drooping invalid, prolonging his life, and, if he be not too far

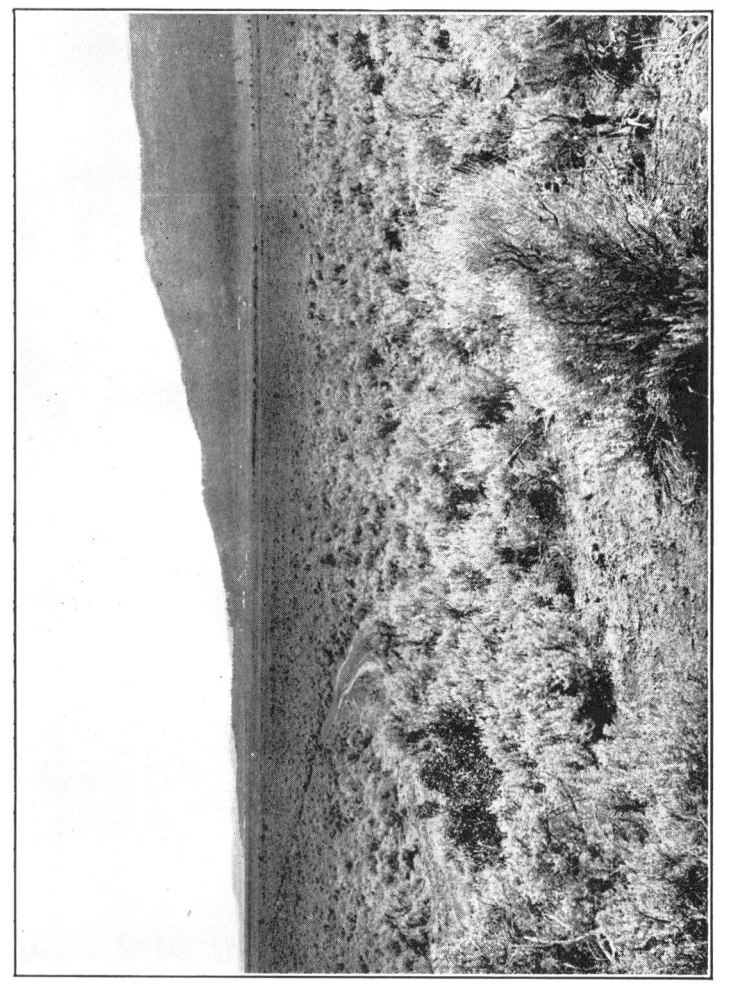

THE SAGEBRUSH DESERT BEFORE RECLAMATION.

gone at the outset, restoring the old vigor to the shattered body. The faces of the permanent sojourners within their influence they paint with the brown badge of health. It is too early as yet to observe the full effect of the climate on the population of the arid West, but sufficient results are apparent to warrant the assertion that these influences will breed a great race.

The element of aridity not only fosters health, but moderates and makes more readily bearable the summer's heat and winter's cold. It is the damp cold that penetrates to the marrow. It is the humid heat that prostrates. To say that a cold of thirty degrees below zero at Helena, in Montana, is felt less than ten degrees above zero in Chicago or New York; or to say that eighty-five degrees above zero in the East is more dangerous to the laborer than one hundred and fifteen degrees at Indio, in the Colorado desert, is to put a severe tax on popular credulity. Nevertheless, both statements are literally true, as all who have experienced the conditions testify.

Science corroborates the story. The United States Weather Bureau has perfected in recent years an instrument to measure the difference between apparent and sensible temperature, which is determined by humidity, or lack of it. The instrument, which consists of a dry and of a wet thermometer, has been in operation at Yuma, in southwestern Arizona, since 1888. Mr. A. Ashenberger, the official observer, reports that the hottest day in that period was July 20, 1892. On that day the dry thermometer registered one hundred and fourteen degrees of apparent heat, and the wet thermometer sixty-nine degrees of sensible heat—a difference of forty-three degrees. The scientific findings are borne out by the every-day testimony of individuals. Sun-strokes in the arid region are practically unknown. The rainless air that sweeps over the arid lands of western America is necessarily dry. It neither breeds diseases nor carries their germs. It is the very breath of health. The lack of moisture, combined with the configuration, forbids the presence of tornadoes, and the Weather Bureau has absolutely no record of such a calamity west of the ninety-seventh meridian.

The superior climate of the arid West is due to fundamental conditions which differ widely from those of eastern America. Viewed from the stand-point of the broader climatic effects, the eastern half of the United States is one wide plain. The moisture-laden winds from lakes and gulf, as from the great ocean itself, meet none but insignificant barriers. But in the Far West the mountains are the supreme factor in the making of the climate. The coast range stands eternal guard along the margin of the sea, while a little farther inland the Sierra Nevada lifts its giant peaks to intercept the clouds which escape the outer barrier and to condense their moisture into snow. Down the centre of the continent, from Canada to Mexico, the Rocky Mountains tower far into the sky, repeating upon the eastern edge of the arid region the process of condensing and storing the winter's rain and holding it against the summer's need. Between the three great primary ranges scores of shorter ones, or isolated mountain groups, reach their long arms into the desert. The dryness, purity, and lightness of the atmosphere are due to this mountain topography, and to the high average altitude throughout the region. It is, then, in the striking character of its climate, springing from these fixed and fundamental conditions, that the great West scores its first superiority over the well-settled states east of the Mississippi river.

But the nation's sanitarium is also the nation's treasure-house. Without the store of precious metals which sleeps in the bosom of the western mountains the American people would be practically dependent on foreign lands for their supply of gold and silver. From this pitiable plight the nation was saved by the wise statesmanship and the great good fortune which brought into the Union the States of Colorado, Utah, and California, of Idaho, Montana, and Nevada, of Washington, Oregon, and Wyoming, and the Territories of New Mexico and Arizona. European nations testify their appreciation of such resources by struggling for the possession of South Africa, a mineral field scarcely worthy to be mentioned in comparison with that of our own great West.

The western half-continent is rich not merely in the precious metals, but in all the raw materials of economic greatness. Its supreme advantage consists in the extraordinary diversity of its resources. In sketching the peculiarities of the several Western States, further on in these pages, the facts will be stated with more detail. In directing attention to the general superiority of these States over their sisters of the East, it is sufficient now to say that they have more water-power than New England; more coal, iron, and oil than Pennsylvania; larger and better forests than Maine and Michigan; and produce better wheat and corn than Illinois and Indiana. The time is rapidly coming when they will produce more and better sugar than Louisiana, and will revolutionize the tanning industry by supplanting oak and hemlock bark with canaigre. With beef and mutton, wool and hides, they already feed and clothe the East. They have finer harbors than Boston and New York, and a sea-coast which faces a greater foreign world.

There is no Eastern State that compares with almost any one of these giant commonwealths of the comparatively unknown West in anything save present development, which includes, of course, population, wealth, and political influence. So emphatic and unmistakable is the superiority with which nature endowed the Far West that it may be said in all seriousness that if the Pilgrim Fathers had landed at San Diego rather than at Plymouth, that half of the country which now contains over ninety per cent. of the total population would be regarded as comparatively worthless. It would have been difficult to settle it to the best advantage. To illustrate: imagine the excitement which would occur if the people of New England should awaken some morning to find themselves in possession of the climate and diversified resources of Colorado, Washington, or California! Even the sane brain which rules the land of steady habits would grow dizzy in the presence of such vast possibilities. And yet Colorado, Washington, and California represent but a small proportion of the country which rests under the wide arch of our western sky.

In briefly reviewing the salient points of difference between the old section and the new, the feature which constitutes at once the most characteristic and the most fundamental advantage of the West has been left for separate treatment. Not until this feature has been con-

BETTER HALF OF THE UNITED STATES

sidered is it possible to appreciate the striking character of the new civilization which will rule the destinies of the western half of the continent, and, very probably, project new and potent influences into the social and political life of the United States as a whole.

CHAPTER IV

THE BLESSING OF ARIDITY

FORTUNATE beyond all other parts of the United States in its climate and in the surpassing wealth of its forests, its quarries and its mines, western America is yet more favored in another element of its physical foundation. This is the substantial aridity which prevails throughout its vast proportions.

The anomaly that its foremost blessing should consist in the fact which gave it a wide-spread reputation for worthlessness is interesting, but unimportant. Nature frequently conceals her raw materials of greatness, alike in men and in countries, until time and opportunity are ripe. In the aridity of the West we shall find the true key to its future institutions. Climate may produce a healthy race, and mineral resources may enrich it, but the natural conditions which determine the character of social and industrial organization, and mould the habits and customs of men, are the potent influences which shape civilization. Hence we shall see that in any just estimate of the relative worth of western resources the fact of aridity must be rated as high above the value of forests and mines as human progress is dearer than money, and as the fate of the race is more momentous than the prosperity of individuals.

TWO YEARS AND FOUR MONTHS PRIOR TO THE TAKING OF THIS PHOTOGRAPH, NEITHER WATER, TREE, NOR MAN EXISTED IN THIS PLACE, IMPERIAL VALLEY, CALIFORNIA.

THE BLESSING OF ARIDITY

The influence of the new environment may readily be illustrated by comparing the conditions which confronted the early settlers of the New England forests and the Illinois prairie, on one hand, and, upon the other, those which the settler met in the deserts around Salt Lake. Except for the temporary need of defence against the Indians, eastern settlers were able to locate their homes without reference to neighbors. They cleared the forest or turned the prairie sod, and were ready to begin. They generally took all the land they could claim under the law, and held much of it out of use for speculation. The greed for land resulted in large farms, and this involved social isolation. The individual acted alone and exclusively for his own benefit. The conditions not only favored, but practically compelled it. Out of this primal germ of our eastern citizenship grew the plant of individual enterprise, which is the conspicuous product of the time. The fruit which it bore was competition, and this has latterly tended towards monopoly.

The conditions which confronted the settler in the deserts of Utah were widely different. There he could not build his home and make his living regardless of his neighbor. Without water to irrigate the rich but arid soil he could not raise a spear of grass nor an ear of corn. Water for irrigation could only be obtained by turning the course of a stream and building canals which must sometimes be cut into the solid walls of the canyon or conducted across chasms in flumes. All this lay beyond the reach of the individual. Thus it was found that the association and organization of men were the price of life and prosperity in the arid West. The alternative was starvation. The plant which grew from this new seed was associative enterprise, and we shall presently see what flower it bore in Utah and other States of the arid region. But it is interesting to first observe that we have encountered in these underlying conditions of the western half-continent principles that are as old as history and as wide as humanity.

The founders of the wonderful civilization of the Netherlands were compelled to deal with conditions which brought into action the same forces as those which are working out interesting results in the arid region of the United States. The Dutch combined and organized their efforts in order to keep the water off their lands, as the Westerners combine and organize to bring the water on. Writing of this aspect of his subject in that enlightening book, *The Puritan in Holland, England, and America,* Mr. Douglass Campbell says:

"The constant struggle for existence, as in all cases when the rewards are great enough to raise men above biting, sordid penury, strengthens the whole race, mentally, morally, and physically. *Labor here has never been selfish and individual. To be effective, it requires organization and direction.* Men learn to work in a body and under leaders. A single man laboring on a dike would accomplish nothing; the whole population must turn out and act together."

Even more interesting and significant is Mr. Campbell's statement of the far-reaching influence, upon the whole economic fabric of the nation, of the co-operative methods taught the founders of Holland by the necessities of their situation and transmitted to their descendants. He says:

"The habits thus engendered extend in all directions.

THE BLESSING OF ARIDITY

Everything is done in corporations" [co-operations?]. "Each trade has its guild, elects its own officers, and manages its own affairs. The people are a vast civic army, subdivided into brigades, regiments, and companies, all accustomed to discipline, learning the first great lesson of life—obedience."

Professor E. W. Hilgard, the distinguished director of the agricultural department of the University of California, has brought this line of reasoning from physical causes to industrial effects into direct application to our subject. In a notable contribution to the *Popular Science Monthly* he says:

"As irrigation means heavy investments of capital or labor, hence the co-operation of many and the construction of permanent works: it necessarily implies the correlative existence of a stable social organization, with protection for property rights, and (in view of the complexity of the problem of proper and equitable distribution of water) a rather advanced appreciation of the need and advantages of co-operative organization."

It was in the course of an effort to account for the singular preference of the founders of the most ancient civilizations for arid lands, rather than for the forested areas which have been the scenes of later development, that Professor Hilgard made this expression of the obvious effects of irrigation on industrial polity. A little further on we shall see other interesting results of his inquiry in this field.

The quality of aridity is thus the most significant among many striking contrasts which mark the western half of the United States—the field for future settlement and development—as fundamentally different

THE CONQUEST OF ARID AMERICA

from the eastern half. Its relation to agriculture is important and interesting, but its relation to a future civilization in a broader sense will be momentous. It is, indeed, a fateful crop, trembling with the hopes of humanity, that is beginning to sprout from the arid soil of the far-western deserts.

The blessing of aridity is again conspicuously illustrated in its remarkable effect upon the soil. The land which the casual traveller, speaking out of the splendid depths of his ignorance and prejudice, condemns as "worthless" and fit only "to hold the earth together," is in reality rich and durable beyond the most favored districts in the humid regions. It is the marvel of every eastern farmer who comes in contact with it. Professor Hilgard sees in this phenomenal fertility the most reasonable explanation of the choice of arid lands by the people foremost in ancient civilization.

It has puzzled the historian to account for the fact that the glories of antiquity sprang from the heart of the desert. The fact itself is, of course, beyond dispute. Egypt, Asia Minor, and Syria, with Palestine, "the land of milk and honey"; Persia, Arabia, and the classic lands of northern India, as well as the countries of the Carthaginians and the Moors, were arid regions. So also were the chosen homes of the Incas in South America, and of the Aztecs and Toltecs in Mexico and our own Southwest, the fame of whose vanished civilizations is reflected in the pages of Prescott and Baldwin. For aught we know to the contrary, these departed nations may have been perfect types of the co-operative commonwealth, and the knack of governing them for the equal benefit of all may be the most precious of the lost

THE BLESSING OF ARIDITY

arts. Among the silent witnesses which have survived the centuries to testify to the engineering skill and the perfection of social organization of those who were swept into oblivion by nameless calamities, are great irrigation canals, portions of which are even yet so true and substantial as to serve the uses of to-day in conjunction with modern works. There are such instances in Arizona.

The accepted explanation of the choice of the arid land by the ancient races is that they sought security against savage enemies, both animal and human, which infested the forest. The theory is purely sentimental and quite inconsistent with the slight but conclusive evidences of their superior intelligence and courage which yet survive. The reasonable explanation of the mystery of ancient civilization is that the arid lands were chosen because they were infinitely better than the humid lands, and because they presented conditions much better suited to the industrial polity of the people and the age.

In searching for the clue of this mystery Professor Hilgard has developed facts which tend to upset other accepted theories. It has long been conceded that certain arid districts are the richest spots on the surface of the globe. "The valley of the Nile," for instance, is a phrase which is everywhere taken as a synonym of extraordinary fertility. The richness and durability of the Nile lands, which have supported for centuries an average population of little more than one and one-half persons to each acre of cultivated soil (a density of settlement which would give Texas a population of over one hundred and sixty millions), has been ascribed to

THE CONQUEST OF ARID AMERICA

the fertilizing quality of the annual deposit of river sediment. The partisans of irrigation have made much of this aspect of the matter, asserting that the artificial application of water is itself a means of fertilization. They have asserted the claim not only where the source of supply, as in the cases of the Rio Grande and the Rio Colorado, is obviously heavily charged with silt held in suspension, but with almost equal ardor in cases where the water flows, a stream of limpid crystal, directly from the mountain-side, or gushes impetuously from the earth in artesian outpourings.

That the famous river Nile does, indeed, leave a thin deposit of rich soil upon each subsidence of its annual flood our California scientist does not, of course, deny. He proves, however, that this layer of new soil is only of the thickness of common cardboard—one-twenty-fifth of an inch—and is equal to only about two good two-horse loads per acre. Three times as much stable manure is the usual dressing for an acre. He truly observes that as the sediment is merely rich soil, thousands of farmers could readily haul and spread such fertilizer upon their land, and would doubtless do so if they could thereby duplicate the phenomenal fertility of the Nile country. He clinches his argument by showing that the neighboring province of Fayoom, in the Libyan Desert, shares the perpetual fertility of the Nile district, though irrigated only with the clear waters of Lake Moeris; that the regur lands of the Deccan, in south-central India, have been phenomenally productive for thousands of years, and that the loess region of China, drained by the headwaters of the Yellow river, have been the granary of China for ages. Like the famous Egyptian

THE BLESSING OF ARIDITY

provinces, the lands referred to in India and China are arid or semi-arid, and, unlike the Nile Valley, they have not been enriched by sedimentry deposits or fertilized by irrigation.

Hence, Professor Hilgard reaches the somewhat sensational conclusion that the extraordinary fertility which, by world-wide acknowledgment, marks the valley of the Nile, is a *quality inherent in aridity itself*. And he maintains his contention thus:

"Soils are formed from rocks by the physical and chemical agencies commonly comprehended in the term *weathering*, which includes both their pulverization and chemical decomposition by atmospheric action. Both actions, but more especially the chemical one, continue in the soil itself; the last named in an accelerated measure, so as to give rise to the farmer's practice of 'fallowing'—that is, leaving the land exposed to the action of the air in a well-tilled but unplanted condition, with a view to increasing the succeeding year's crop by the additional amount of plant-food rendered available, during the fallow, from the soil itself.

"This weathering process is accompanied by the formation of new compounds out of the minerals originally composing the rock. Some of these, such as zeolites and clay, are insoluble in water, and therefore remain in the soil, forming a reserve of plant-food that may be drawn upon gradually by plants; while another portion, containing especially the compounds of the alkalies—potash and soda—are easily soluble in water. Where the rainfall is abundant these soluble substances are currently carried into the country drainage, and through the rivers into the ocean. Among these are potash, lime, magnesia, sulphuric and a trifle of phosphoric acids. Where, on the contrary, the rainfall is insufficient to carry the soluble compounds formed in the weathering of the soil-mass into the country drainage, those compounds must of necessity remain and accumulate in the soil."

All this is perfectly comprehensible, even to the lay mind. The valuable ingredients of the soil which are soluble have been washed out of the land in humid regions, like our eastern States, by the rains of centuries. On the other hand, these elements have been accumulating in the arid soil of the West during the same centuries. They lie there now like an inexhaustible bank account on which the plant-life of the future may draw at will without danger of protest. The process which created this rich soil goes on repeating itself—recreating the soil season after season. The same is true, of course, in the arid and semi-arid regions of Egypt, India, China, and all other localities that enjoy the inestimable blessing of aridity.

Professor Hilgard's conclusions are the result of patient investigation. They are based on more than one thousand analyses of the soils of the arid and the humid regions of the United States—of the West and the East. These analyses demonstrated the following astounding fact: That the soils of the arid regions lying west of the one hundredth meridian, when compared with the soils of the humid region lying east of the Mississippi river, contain on the average three times as much potash, six times as much magnesia, and fourteen times as much lime. This is the scientific explanation of the superior productiveness of the arid regions of the

THE BLESSING OF ARIDITY

West, which every intelligent observer has noted and marvelled to behold.

The people of the Blue Grass Region of Kentucky and of other favored localities have repeated from generation to generation the boast that "a limestone country is always a rich country." Professor Hilgard has demonstrated that the average arid soil is equal to the most phenomenal soil of the East, while the soil of the arid West as a whole is beyond comparison with that of the humid East as a whole. He coins the maxim, "Arid countries are always rich countries when irrigated," and the phrase does scant justice to the subject. It only remains to add that Professor Hilgard is recognized as the foremost expert on soils in the West, and one of the first men in his profession in the United States. No one will question the weight of his views, for they coincide alike with common-sense and with world-wide experience through the centuries. It cannot, therefore, be doubted that the agricultural foundation of the Far West, as it relates to the soil, is incomparably better than any other part of the continent.

While science has thus furnished a lucid explanation of the universal fertility of arid lands, it would be unfair to draw the conclusion that the claims which have been made concerning the rare fertilizing qualities of certain western rivers are entirely unfounded. Nearly all of the rivers in the West carry more or less rich silt, due to the fact that they flow through treeless regions, where the soil is swept into the stream by winds and sudden torrents. Eastern rivers are, as a rule, much clearer, because they flow through forests and cultivated fields. The waters of the Colorado river gather an enormous quantity of fertilizing matter in their long journey from the mountains of Wyoming to the Gulf of California. There is no guesswork in this instance. The scientific men of the University of Arizona, at Tucson, have made patient experiments, extending over many months of time, to determine the actual commercial value of the fertilizer contained in these waters and precipitated on the land in the process of irrigation. Basing their computation upon the use of thirty-six acre-inches of this water, they find that the fertilizing material so applied would cost, if purchased in the market, the sum of nine dollars and seven cents per acre. Where such conditions prevail cultivation can never impoverish, but actually enriches, the fortunate soil. But we have yet to mention the chief blessing of aridity. This is the fact that it compels the use of irrigation.

And irrigation is a miracle!

CHAPTER V

THE MIRACLE OF IRRIGATION

THE beauty of Damascus is the theme of poets. Speaking of this ancient capital an anonymous writer remarks that "the cause of its importance as a city in all the ages is easily seen as you approach it from the south. Miles before you see the mosques of the modern city the fountains of a copious and perennial stream spring from among the rocks and brushwood at the base of the Anti-Lebanon, creating a wide area about them, rich with prolific vegetation." He continues:

"These are the 'streams of Lebanon,' which are poetically spoken of in the Songs of Solomon, and the 'rivers of Damascus,' which Naaman, not unnaturally, preferred to all the 'waters of Israel.' This stream, with its many branches, is the inestimable treasure of Damascus. While the desert is a fortification round Damascus, the river, where the habitations of men must always have been gathered, as along the Nile, is its life.

"The city, which is situated in a wilderness of gardens of flowers and fruits, has rushing through its streets the limpid and refreshing current; nearly every dwelling has its fountain, and at night the lights are seen flashing on the waters that dash along from their mountain home. As you first view the city from one of the overhanging

THE CONQUEST OF ARID AMERICA

ridges you are prepared to excuse the Mohammedans for calling it the earthly paradise. Around the marble minarets, the glittering domes, and the white buildings, shining with ivory softness, a maze of bloom and fruitage—where olive and pomegranate, orange and apricot, plum and walnut, mingle their varied tints of green—is presented to the sight, in striking contrast to the miles of barren desert over which you have just ridden."

This is the miracle of irrigation in the Syrian desert. It is no more miraculous in that far-eastern country than in our own West. Nor is Damascus more beautiful than Denver, Salt Lake City, or than any one of a score of modern towns in California. But because Damascus is ancient and historic, and looks down on mankind from the biblical past, it possesses a degree of interest with which it is difficult to invest much better and more important places of our own country and our own time. It is well, then, to remember that not only the beauty of Damascus, but the glories of the Garden of Eden itself, were products of irrigation. "A river went out of Eden to water the Garden," says the Bible story.

No consideration of the subject can be appreciative when it starts with the narrow view that irrigation is merely an adjunct to agriculture. It is a social and industrial factor, in a much broader sense. It not only makes it possible for a civilization to rise and flourish in the midst of desolate wastes; it shapes and colors that civilization after its own peculiar design. It is not merely the life-blood of the field, but the source of institutions. These wider and more subtle influences are difficult to define in abstract terms, but we may trace them clearly through the history of various commu-

THE CACTUS DESERT TRANSFORMED INTO PARK, AT RIVERSIDE, CAL.

THE MIRACLE OF IRRIGATION

nities which have grown up in conformity with these conditions.

The essence of the industrial life which springs from irrigation is its democracy. The first great law which irrigation lays down is this : There shall be no monopoly of land. This edict it enforces by the remorseless operation of its own economy. Canals must be built before water can be conducted upon the land. This entails expense, either of money or of labor. What is expensive cannot be had for naught. Where water is the foundation of prosperity it becomes a precious thing, to be neither cheaply acquired not wantonly wasted. Like a city's provisions in a siege, it is a thing to be carefully husbanded, to be fairly distributed according to men's needs, to be wisely expended by those who receive it. For these reasons men cannot acquire as much irrigated land, even from the public domain, as they could acquire where irrigation was unnecessary. It is not only more difficult to acquire in large bodies, but yet more difficult to retain. A large farm under irrigation is a misfortune; a great farm, a calamity. Only the small farm pays. But this small farm blesses its proprietor with industrial independence and crowns him with social equality. That is democracy.

Industrial independence is, in simplest terms, the guarantee of subsistence from one's own labors. It is the ability to earn a living under conditions which admit of the smallest possible element of doubt with the least possible dependence upon others. Irrigation fully satisfies this definition.

The canal is an insurance policy against loss of crops by drought, while aridity is a substantial guarantee

THE CONQUEST OF ARID AMERICA

against injury by flood. Of all the advantages of irrigation, this is the most obvious. Scarcely less so, however, is its compelling power in the matter of production. Probably there is no spot of land in the United States where the average crop raised by dependence upon rainfall might not be doubled by intelligent irrigation. The rich soils of the arid region produce from four to ten times as largely with irrigation as the soil of the humid region without it. As the measure of value is not area, but productive capacity, twenty acres in the Far West should equal one hundred acres elsewhere. Such is the actual fact.

A little further on we shall see that not merely the quantity of crops, but their quality as well, responds to the influence of irrigation. We shall see how this art favors the production of the wide diversity of products required for a generous living. Certainty, abundance, variety—all this upon an area so small as to be within the control of a single family through its own labor—are the elements which compose industrial independence under irrigation. The conditions which prevail where irrigation is not necessary—large farms, hired labor, a strong tendency to the single crop—are here reversed. Intensive cultivation and diversified production are inseparably related to irrigation. These constitute a system of industry the fruit of which is a class of small landed proprietors resting upon a foundation of economic independence.

This is the miracle of irrigation on its industrial side.

As a factor in the social life of the civilization it creates, irrigation is no less influential and beneficent. Compared with the familiar conditions of country life

THE MIRACLE OF IRRIGATION

which we have known in the East and central West, the change which irrigation brings amounts to a revolution. The bane of rural life is its loneliness. Even food, shelter, and provision for old age do not furnish protection against social discontent where the conditions deny the advantages which flow from human association. Better a servant in the town than a proprietor in the country!—such has been the verdict of recent generations who have grown up on the farm and left it to seek satisfaction for their social instincts in the life of the town. The starvation of the soul is almost as real as the starvation of the body.

Irrigation compels the adoption of the small-farm unit. This is the germ of new social possibilities, and we shall see to what extent they have already been realized as we proceed. During the first and second eras of colonization in this country the favorite size for a farm was about four hundred acres, of which from a fourth to a half was gradually cleared and the rest retained in woodland. The Mississippi Valley was settled mostly in quarter-sections, containing one hundred and sixty acres each. The productive capacity of land is so largely increased by irrigation, and the amount which one family can cultivate by its own labor consequently so much reduced, that the small-farm unit is a practical necessity in the arid region.

Where settlement has been carried out upon the most enlightened lines irrigated farms range from five to twenty acres upon the average, rarely exceeding forty acres at the maximum. It is perfectly obvious, of course, that a twenty-acre unit means that neighbors will be eight times as numerous as in a country settled

THE CONQUEST OF ARID AMERICA

up in quarter-sections—that where farms are ten acres in size neighbors will be multiplied by sixteen. Thus in its most elementary aspect the society of the arid region differs materially from that of a country of large farms. Eight or sixteen families upon a quarter-section are much better than no neighbors at all, but irrigation goes further than this in revolutionizing the social side of rural life.

A very-small-farm unit makes it possible for those who till the soil to live in the town. The farm village, or home centre, is a well-established feature of life in Arid America, and a feature which is destined to enjoy wide and rapid extension. Each four or five thousand acres of cultivated land will sustain a thrifty and beautiful hamlet, where all the people may live close together and enjoy most of the social and educational advantages within the reach of the best eastern town. Their children will have kindergartens as well as schools, and public libraries and reading-rooms as well as churches. The farm village, lighted by electricity, furnished with domestic water through pipes, served with free postal delivery, and supplied with its own daily newspapers at morning and evening, has already been realized in Arid America. The great cities of the western valleys will not be cities in the old sense, but a long series of beautiful villages, connected by lines of electric motors, which will move their products and people from place to place. In this scene of intensely cultivated land, rich with its bloom and fruitage, with its spires and roofs, and with its carpets of green and gold stretching away to the mountains, it will be difficult for the beholder to say where the town ends and the country begins.

FOUR MONTHS' OLD RANCH, NORTH YAKIMA, WASHINGTON.
(Courtesy of The World's Work.)

THE MIRACLE OF IRRIGATION

This is the miracle of irrigation upon its social side.

Irrigation is the foundation of truly scientific agriculture. Tilling the soil by dependence upon rainfall is, by comparison, like a stage-coach to the railroad, like the tallow dip to the electric light. The perfect conditions for scientific agriculture would be presented by a place where it never rained, but where a system of irrigation furnished a never-failing water supply which could be adjusted to the varying needs of different plants. It is difficult for those who have been in the habit of thinking of irrigation as merely a substitute for rain to grasp the truth that precisely the contrary is the case. Rain is the poor dependence of those who cannot obtain the advantages of irrigation. The western farmer who has learned to irrigate thinks it would be quite as illogical for him to leave the watering of his potato-patch to the caprice of the clouds as for the housewife to defer her wash-day until she could catch rain-water in her tubs.

The supreme advantage of irrigation consists not more in the fact that it assures moisture regardless of the weather than in the fact that it makes it possible to apply that moisture just when and just where it is needed. For instance, on some cloudless day the strawberry-patch looks thirsty and cries for water through the unmistakable language of its leaves. In the Atlantic States it probably would not rain that day, such is the perversity of nature, but if it did it would rain alike on the just and unjust—on the strawberries, which would be benefited by it, and on the sugar-beets, which crave only the uninterrupted sunshine that they may pack their tiny cells with saccharine matter. In the arid region there is practically no rain during the growing season. Thus the

THE CONQUEST OF ARID AMERICA

scientific farmer sends the water from his canal through the little furrows which divide the lines of strawberry plants, but permits the water to go singing past his field of beets.

Plants and trees require moisture as well as sunshine and soil, and for three reasons: first, that the tiny roots may extract the chemical qualities from the soil; then, that there may be sap and juice; finally, that there may be moisture to evaporate or transpire from the leaves. But while all plant-life requires moisture, all kinds of it do not require the same amount, nor do they desire to receive it at the same time and in the same manner. Just as the skilful teacher studies the individualities of fifty different boys, endeavoring to discover how he may most wisely vary his methods to obtain the best results from each, so the scientific farmer studies his fifty different plants or trees and adjusts his artificial "rainfall" in the way which will produce the highest outcome. With the aid of colleges, experimental farms, and county institutes, wonderful progress has been made along these lines in recent years. This progress will continue until the agriculture and horticulture practised on the little farms of Arid America shall match the marvellous results won by research and inventive genius in every other field of human endeavor.

This is the miracle of irrigation upon its scientific side.*

* For full explanation of practical methods of irrigation, see Appendix.

Part Second

REAL UTOPIAS OF THE ARID WEST

"At every new stage of the history of the American settlement, we are afresh reminded that colonies are planted by the uneasy. The discontent that comes from poverty and financial reverse, that which is born of political unrest, and that which has no other cause than feverish thirst for novelty and hazardous adventure, had each a share in impelling Englishmen to emigrate. But in the seventeenth century religion was the dominant concern—one might almost say the dominant passion—of the English race, and it supplied much the most efficient motive to colonization. Not only did it propel men to America, but it acted as a distributing force on this side of the sea, producing secondary colonies by expelling from a new plantation the discontented and the persecuted to make fresh breaks in the wilderness for new settlements."—EDWARD EGGLESTON, *Beginners of a Nation.*

CHAPTER I

THE MORMON COMMONWEALTH

To study the human side of things in the arid region of the Far West, we must begin with the Mormon Commonwealth of Utah. This is true for a number of excellent reasons. We find here the earliest development of any consequence. Although irrigation is older than history, it was never practised upon any considerable scale by Anglo-Saxons until the Mormon pioneers turned the waters of City Creek upon the alkaline soil of Salt Lake Valley in the summer of 1847.

In Utah, almost alone of the far-western States, settlement began with home-making pure and simple. Irrigation was the primal and single industry until a commonwealth had been established. In California, in Colorado, in Nevada, in Idaho, and in Montana, mining, rather than agriculture, was the motive which induced the original settlement by Americans, and irrigation grew up only as an adjunct to the mining camp. In Wyoming, and in a less degree elsewhere, stock-raising was the first pursuit and irrigation was used merely to flood the bottom land and grow crops of coarse, wild hay for the winter feeding of cattle. In Washington and Oregon the first settlements were made along the humid coast region, and the arid parts of those States were settled, in such measure as they have been settled at all, by the overflow of those original currents of population. But in Utah the motive was home-building, and the pursuit was agriculture for its own sake.

Furthermore, we find in Utah, and nowhere else, an entire and distinct people, who have grown up under one strong and simple industrial system, and have brought that system to its logical results. This experience covers half a century, and cannot be objected to on the ground that it is an experiment, the results of which remain to be demonstrated.

Finally, partly because of these several reasons and partly because the Mormon fugitives possessed no capital except their leader's brains and their own hard hands, the economic institutions of Utah are the natural outgrowth of the conditions of an arid land. Utah is the product of its environment. As we study it we shall see the economic tendencies underlying and shaping the industrial life of all communities which find their life-current in the irrigation canal and are surrounded by the rich and varied, but wholly undeveloped, resources of our far-western country. It is for these reasons that the Mormon Commonwealth suggests itself irresistibly as the starting-point of any proper study of our subject.

What did the pioneers have to start with? What have they accomplished in fifty years? How did they do it? In the answers to these questions we may find a flood of light for the future of the West, but only upon condition that the answers be sought in a spirit of perfect candor and without prejudice either in favor of or against the interesting people of the Utah mountains.

On July 24, 1847, the Mormon caravan emerged from

the mouth of Emigration Canyon into the valley of the Great Salt Lake. It was a beautiful picture that greeted the eyes of the fugitives as they rested here to enjoy the shade of the cottonwoods and listen to the music of the mountain torrent and the birds. Out of the chill air of the higher altitudes, out of the dark shadows of the picturesque chasm, they had come by a sudden turn face to face with a broad, sunlit valley, which sloped gently away to the shore of an inland sea. On the east, the Wasatch mountains reared their brown and rifted barriers until their summits were lost in a crown of eternal snows. To the south and west the Oquirrhs marshalled their peaks into the waters of the lake. Below them, valley and lake; around them on every side, mountains and more mountains; over them, the impalpable sky—this was the vision which burst suddenly upon the tired eyes of the pilgrims.

When they had proceeded a little farther they caught sight of a large fresh lake some miles to the south, emptying its surplus waters into an inland sea through a slender river, which shone like a ribbon of silver. The comparison suggested by these strange conditions might have occurred to a duller mind than that of Brigham Young, who felt that he was a Moses leading a new tribe of Israel to a new promised land. The fresh lake was the sea of Tiberius; the salt one, the Dead Sea; the river was, of course, the Jordan. This, then, was the new Palestine, and here the leader and his followers would build the new Jerusalem! Advancing a few miles into the valley, and halting near the banks of a roaring brook, Brigham Young struck his staff upon the ground and exclaimed, "Here we will rear our temple in holiness to the Lord." It is above this spot that Sculptor Dallin's graceful figure of the Angel Moroni now looks down from a stately pile of Utah granite, reared at a cost of forty years' labor and six million dollars.

The pioneers possessed very little cash capital when they arrived in the valley which was to be the heart of a future commonwealth. This was not a serious misfortune, since there was little that money would buy in Utah at that time, or anywhere within one thousand miles east, west, north, or south. They had located at almost the exact geographical centre of that great arid region whose modern agricultural era they were destined to inaugurate. Surrounded by extraordinary wealth, there was but one thing which could pass current as a medium of exchange in this primeval wilderness. This one thing was labor, and the free and unlimited coinage of labor has been the cardinal doctrine in Utah's economic faith from the beginning down to the present hour. Besides their willing industry, the Mormons had brought with them the contents of seventy-two wagons, about one hundred horses, less than half as many mules and oxen, nineteen cows and a few chicken. It was with this capital that they began the making of Utah. But at the very threshold of their life in a new country they were confronted by something utterly strange to them in the conditions of agriculture.

First of the Anglo-Saxon race, the Mormons encountered the problem of aridity, and discovered that its successful solution was the price of existence. Brigham Young had lived in Vermont, Ohio, Missouri, and Illinois. Neither he nor any of his followers had ever seen a country where the rainfall did not suffice for agri-

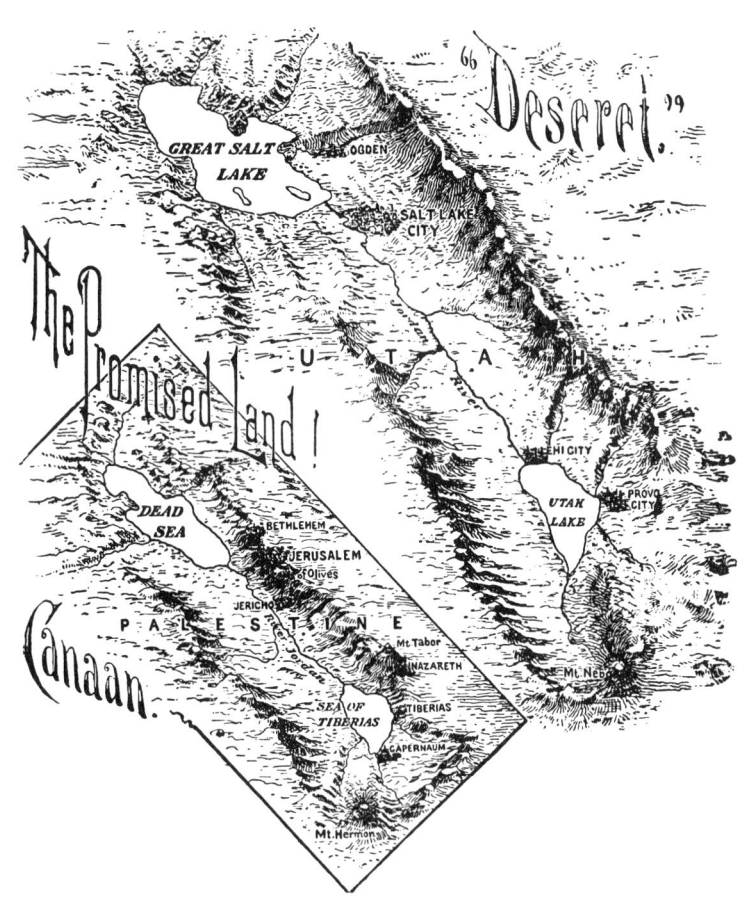

MAP SHOWING THE STRIKING SIMILARITY BETWEEN PALESTINE
AND SALT LAKE VALLEY, UTAH

(By courtesy of the Rio Grande Western R. R.)

culture, nor ever read of one save in the Bible. But they quickly learned that they had staked their whole future upon a region which could not produce a spear of tame grass, an ear of corn, nor a kernel of wheat without skilful irrigation. Of the art of irrigation they were utterly ignorant. But the need of beginning a planting was urgent and pressing, for their slender stock of provisions would not long protect them from starvation.

It was this emergency which produced the first irrigation canal ever built by white men in the United States. Mormons are prone to believe that the suggestion of this work was a revelation from God to the head of the Church. Other traditions ascribe it to the advice of friendly Indians; to the example of the Mexicans; to the shrewd intuition with which the leader had met all the trials encountered in the course of his adventurous pilgrimage. Whatever the source of the inspiration, he quickly set his men at work to divert the waters of City Creek through a rude ditch and to prepare the ground for Utah's first farm. These crystal waters now furnish the domestic supply for a city of sixty thousand inhabitants. The late President Wilford Woodruff, who was one of the party assigned to the work of digging the first canal, related that when the water was turned out upon the desert the soil was so hard that the point of a plough would scarcely penetrate it. There was also much white alkali on the surface. It was, therefore, with no absolute conviction of success that the pioneers planted the very last of their stock of potatoes and awaited the result of the experiment. The crop prospered in spite of all obstacles, and demonstrated that a living could be wrung from the forbidding soil of the desert when men should learn to adapt their industry to the conditions.

Such was the humble beginning of modern agriculture in Arid America. The success of this desperate expedient to preserve the existence of a fugitive people in the vast solitude has made Utah our classic land of irrigation, and given the Mormons their just claim as the pioneer irrigators of the United States. It was not, however, until they survived other hardships, including the devastation of their first crops by swarms of crickets, that the hardy settlers were able to celebrate a genuine harvest-home, and to feel that the ground was at last firm beneath their feet. Then began that long era of material prosperity which will never cease until the people depart from the industrial system established by Brigham Young.

It is this industrial system which makes the Mormons well worthy of study at this time. Nothing just like it exists elsewhere upon any considerable scale, yet its leading principles are certainly capable of general application. Good Mormons regard the system, like all their blessings, as a direct revelation of God. Many others consider it the intellectual product of a great man's brain. But when it is studied in connection with Mormon colonization, it is plain that the system was born of the necessities of the place and time—that it is the legitimate product of the peculiar environment of the arid region. The forces that have made the civilization of Utah will make the civilization of western America. It is in this view of the matter that we shall find our justification for a careful study of the Mormon structure of industry and society.

THE MORMON COMMONWEALTH

The economic life of Utah is founded on the general ownership of land. Speaking broadly, all are proprietors, none are tenants. Land monopoly was discountenanced from the beginning. All were encouraged to take so much land as they could apply to a beneficial purpose. None were permitted to secure land merely to hold it out of use for speculation. The corner-stone of the system was industrialism—the theory that all should work for what they were to have, and that all should have what they had worked for. In order to realize this result, it was necessary that each family should own as much land as it could use to advantage, and no more.

The adoption of this principle was plainly due to the peculiar conditions which the leader saw about him. He instantly realized that value resided in water rather than in land; that there was much more land than water; that water could only be conserved and distributed at great expense.

If he had settled in a land of abundant rainfall it is improbable that he would have set such severe limitations upon the amount of land which individuals should acquire. In that case he would, perhaps, have thought it well for his people to take all the land they could possibly obtain under the law, and thus enjoy large speculative possibilities. But if he had pursued this policy in Utah he could not have accommodated the thousands whom he expected to follow him in the early future. He thus found it imperatively necessary to restrict the amount of land which each family should acquire, suiting it to their actual needs. He came from a country which had been settled in farms ranging from two hundred to four hundred acres in size. The reduction in the farm unit

THE CONQUEST OF ARID AMERICA

which he now proposed must have seemed nothing less than startling to his followers. It is plain that in proposing such an innovation he not only comprehended the social necessities of the situation, but anticipated, with remarkable foresight, the possibilities of intensive agriculture by means of irrigation.

The first settlement which he planned was, of course, Salt Lake City and its neighborhood. This became the model of all future colonies. It was laid out in such a way as to secure an equitable division of land values among all the inhabitants.

The city blocks consist of ten acres each, divided into eight lots of one and a quarter acres. These lots were assigned to professional and business men. Next there was a tier of five-acre lots. These were assigned to mechanics. Then there were tiers of ten-acre and of twenty-acre lots. These went to farmers, according to the size of their families. Under this arrangement every colonist was a small landed proprietor, owning a certain amount of irrigated soil from which he could readily produce the necessities of life. The division of land values was remarkably even, for what one man lacked in area of his possessions he gained in location. The small lots were close to the centre of business; the large lots more remote from that centre. As the place grew in course of years from an emigrants' camp to a populous city, with paved streets, domestic water, electric lights, and railways, the inevitable rise in values was distributed with remarkably even hand. Not a single family or individual failed to share in the great fund of "unearned increment" which arose from increasing population and growing public improvements.

THE MORMON COMMONWEALTH

This principle of universal land ownership, and of careful division according to location and of differing needs of various classes, has been followed throughout the Mormon settlements of Utah and surrounding States, and is being duplicated to-day in the latest colonies established by this people.

It is important to note that the Mormon land system rested on individual proprietorship. There never was any attempt at community ownership. The unit of the State was the family and the home. But the moment we pass from the sphere of individual labor we encounter another principle, which was always applied, though not always by the same methods, to public utilities. This was the principle of public ownership and control.

If the Mormon leaders had desired to organize their industrial life in a way to make large private fortunes for themselves, no single item in the list of Utah's resources would have offered a better chance for speculation than the water supply. It was perfectly feasible under the law for private individuals or companies to appropriate these waters, construct canals, sell water rights, and collect annual rental. By adopting this method, which widely prevails in other western States, they could have laid every field, orchard, and garden—every individual and family—under tribute to them and their descendants forever. Neither in law nor in practice, at that time, was it any more a moral and economic wrong to appropriate privately and hold against the public the natural wealth of the streams than it was to do the same with the natural wealth of the mineral belts on government land.

Probably the Mormons owed their escape from the

misfortune of private irrigation works mostly to the fact that this feature of their institutions was established when none of their people possessed sufficient private capital to engage in costly enterprises. They started upon a basis of equality, for they were equally poor. They could buy water rights only with their labor. This labor they applied in co-operation, and canal stock was issued to each man in proportion to the amount of work he had contributed to its construction. This in turn was determined by the amount of land he owned, the owner of twenty acres doing just twice as much work as the owner of ten. Here we see the influence of aridity not only favoring, but compelling, the adoption of the principle of associative enterprise, as mentioned in a previous chapter. But before discussing the wider results of this influence in the life of Utah, it is important to observe the characteristic forms of agriculture which grew out of these new conditions.

We have seen that Brigham Young had made twenty acres the maximum size of farms in the Salt Lake settlement. He now proceeded to lay down a philosophy very different from that which prevailed on the large farms of the wheat and corn country whence he had come. He urged that each family should realize the nearest possible approach to absolute industrial independence within the boundaries of its own small farm. His sermons in the tabernacle dealt less in theology than in worldly common-sense. The result is an agricultural system peculiar to Utah.

Just as we have the cotton-belt in Texas, the corn-belt in Nebraska, the wheat-belt in Dakota, and the orange-district in California, so in Utah we have the land of the diversified farm. This is the first and one of the most precious fruits of the industrialism which had been so deeply rooted in the plan of general land ownership.

Much of the misfortune which the settlers of the Mississippi Valley have endured during the past decade is due to the fact that their industrial system was founded on the speculative instinct. They acquired large farms, because they hoped to get rich out of the rise in land. They engaged in the production of single crops, because they were gambling on the hope of great prices for these staples. They mortgaged their homesteads to make costly improvements, because they had the utmost faith in future big prices for the land and its product. It is very easy to comprehend the virtues of Utah industrialism when we may make use of a Texas cotton plantation or a Dakota wheat farm for a background. In the one case we see the little unmortgaged farm, its crops insured by irrigation, systematically producing a variety of things required for the family consumption. A generous living is within the control of the proprietor of such a home. In the other case we see the single crop exposed to the mercy of the weather and the markets, its owner employing many hired hands, and going to the town to buy with cash nearly all that is necessary to feed his family and laborers.

The Utah system was clearly the outgrowth of the peculiar conditions with which the Mormons dealt. They were so far removed from all centres of production as to make self-sufficiency an imperative condition of existence. Hence they were taught the gospel of industrial independence in its purest and most primitive form. And self-sufficiency is the most striking characteristic of their

THE CONQUEST OF ARID AMERICA

civilization to-day. Wars and panics have swept the country since the pioneers built their homes in Salt Lake Valley, but they and theirs have not gone hungry for a day or an hour. Nor need they do so while water runs down hill and mother earth yields her increase.

The conquest of Utah began with the establishment of agriculture, which is everywhere the foundation of civilization. Brigham Young realized, as the American people may well do to-day, that there can be no prosperity if agriculture languishes. He realized that whatever the Mormon people might have in the future—whatever of factories, stores, and banks, whatever of churches temples, and tabernacles—must come primarily from the surplus profits of the soil.

As soon as his people had been supplied with food and shelter, he turned his attention to the development of a broader industrial life. Workshops, stores, and banks were necessary to furnish facilities for manufacture, distribution, and exchange. All these enterprises were undertaken in a co-operative way under the familiar forms of the joint-stock company. Those who were unwilling to engage in them upon these terms generally left the church and set up for themselves. At the beginning there was no capital for such undertakings except the capital which resided in every man's land and labor—no wealth but the commonwealth. As all had started on a basis of equality, so all were given an equal chance to participate in the new industrial, mercantile, and banking enterprises of the Territory. When a factory or store was to be started subscription papers were circulated and everybody urged to take some of the stock. Payments were made sometimes in cash, more often in products, not infrequently in labor. Of one thing there has never been a scarcity in Utah—this is the chance to work. And labor has always been exchangeable there for other commodities, including bank and mercantile stock. Otherwise it would not have been possible to have secured anything like the wide distribution of these stocks which now prevails.

In the early years the industries were of a crude sort. Everything had to be hauled in ox-teams over a thousand miles of deserts, plains, and mountains. The people used almost no money in their daily transactions. As a medium of exchange they had printed slips of paper known as "tithing-house scrip." This answered every purpose of exchange money, while the prices of commodities were regulated by the standard of values which prevailed elsewhere. But while the local scrip did very well for all home purposes, it did not enable the people to purchase the supplies of machinery which they needed from abroad. The process of equipping their factories was, therefore, necessarily slow, but they rapidly developed an army of skilled artisans, which was constantly augmented by immigration. But even without assistance from the great world which lay so far beyond the borders of their own valleys marvellous progress was achieved in the arts and industries.

Brigham Young was strenuously opposed to the development of the mines by his people, believing that what they might gain in wealth from that source would be much more than offset by the demoralization which would come to his industrial forces with the rise of the speculative spirit. Above all other virtues he placed that of sober industry, earning its bread in the sweat of men's faces. That the mines would some day be worked by " Gentiles " he had no doubt, and he rightly calculated that his own people would enjoy more prosperity by feeding the miners than by working the mines. A few of the many millions afterwards taken from the mountains around Salt Lake would have facilitated the growth and equipment of the Mormon industries immensely during the early years. But time and patience accomplished in the end all — perhaps more than an abundance of original capital might have done. Nearly all the industries essential to a complex and symmetrical business economy have been established for many years. Every important settlement has its co-operative store and bank. From the great beet-sugar factory at Lehi down to the smallest mercantile enterprise in the smallest hamlet, the business is owned by a multitude of stockholders.

The capital represents the surplus profits of the many. The system bears no likeness to Socialism. Nothing is owned by virtue of citizenship nor of membership in the church. No one owns a dollar's worth of stock who has not earned and paid for it. The system is nothing but the joint-stock company with what may be called a generous and friendly interpretation. That is to say, it is really desired that everybody shall have an interest, and that all shall share the benefits. It should not be understood by any means that all have an equal ownership in these various enterprises, for the Mormon system has not resulted in making men equally successful. All have had an equal chance however, and the weak have been watched over and assisted by the strong. Indeed, this latter is one of the few good results to be credited to the exercise of church authority in secular affairs.

It would be quite impracticable to attempt to follow the history of any considerable number of the many co-operative enterprises of Utah. Neither are figures available for a satisfactory generalization of results. But the whole system is typified in the experience of one monumental enterprise—Zion's Co-operative Mercantile Institution. This great house is in a sense the mother and the model of all the Mormon stores in Utah and its surrounding States. Mr. Thomas G. Webber, the successful superintendent of "Z. C. M. I.," as it is familiarly called, describes the history of the enterprise as follows:

"The Institution was organized October 16, 1868; commenced business March 1, 1869; was incorporated for twenty-five years from October 5, 1870, and the capital was then $220,000. It was reincorporated for fifty years September 30, 1895, with a capital stock of $1,077,000.

"During the life of our first incorporation period we have sold $76,352,686 worth of merchandise, and paid to the railroad and express companies for freight $6,908,630.

"We have paid out in cash dividends $1,990,943.55, and in stock dividends $414,944.77. During the panic in 1873, for prudential reasons, we passed our dividend, and continued to do so until 1877, but during the whole of the period we have been in business, some twenty-seven years, we have paid to our stockholders an average dividend of nine and one-third per cent. for each and every year, or two hundred and forty-three per cent. in all; $1,000 invested in our capital stock on the 1st of

March, 1869, at the end of September, 1895, when our incorporation ran out, had accumulated to $2,014.30, and in addition to this we have paid upon this $1,000 in cash dividends the sum of $4,218.05.

"We have turned out in our manufacturing departments boots and shoes to the value of $2,053,294.43, and in our duck clothing and shirt factory upwards of $80,000 worth. Last year (1895) it was an off-year with our manufacturing departments, but we turned out 75,400 pairs of boots and shoes, and 15,648 dozen overalls, shirts, etc."

This is the history of Utah's largest co-operative undertaking. It is a history which no friend of co-operative effort will blush to read, for it proves that a great business can be as successfully administered in the interest of the many as in the interest of a few. The latest and largest of the Mormon industrial enterprises is the beet-sugar factory, owned by seven hundred stockholders, which in 1895 produced considerably more than 700,000,000 pounds of sugar and paid a cash dividend of ten per cent. Its later dividends are much larger. It also furnished a profitable market for the products of many irrigated fields.

While the most satisfactory results of co-operative enterprise have been obtained in the last two decades, much was achieved in the early day. As early as 1850, when Salt Lake Valley had been settled less than three years, the industrial products amounted to only a little less than three hundred thousand dollars. Ten years later they had mounted nearly to the million mark, and in 1870 they considerably exceeded two and a quarter millions. In 1895 the total was close to six millions. The

THE MORMON COMMONWEALTH

growth of these hard-won industries has naturally fostered a feeling of intense loyalty to home products. Foreign goods are not a badge of honor. The Utah man wears Utah clothes, made in Utah factories, from wool sheared from the back of Utah sheep, with the same pride that a New York man wears a London hat and a New York woman a Paris gown.

Let us look now at the broader results of the Mormon labor in the wilderness. The church historian, Mr. A. Milton Musser, has made a careful estimate of the financial results which may fairly be credited to the irrigation industry in Utah. In doing so he communicated with church leaders throughout the State and compiled the results of his correspondence with the utmost care. The statement is given just as he prepared it, without attempt to discuss it in detail. To fully comprehend it however, the reader must remember that the Mormons began in poverty, having almost nothing to invest except the labor of their hands and brains, and that all they have expended in a period of nearly fifty years for all classes of improvements—from the first shanty to the last turret of the last temple—came primarily from the soil. Here is the balance-sheet of the Mormon people as Mr. Musser prepared it:

Cost of establishing the 10,000 farms ($187.50 per farm per annum)...................	$75,000,000
Cost of making irrigation canals and ditches ($37.50 per farm per annum)...........	15,000,000
Cost of irrigating 10,000 farms and gardens ($24.00 each per annum)...............	9,600,000
Building factories.......................	5,000,000
Building temples........................	8,000,000

Building churches and schools.............	$4,000,000
Cost of missionary work..................	10,000,000
Cost of immigrating and sustaining the poor	8,000,000
Living of the farmers ($875 to each family per annum).........................	350,000,000
Cost of roads and bridges in mountains and valleys.............................	4,000,000
Cost of Indian wars, building forts, stockades, breaking up settlements, etc...........	5,000,000
Cost of feeding and clothing Indians and establishing Indian missions, farms, schools, etc.........................	2,000,000
Cost of resisting the invasion of the army of 1857, and of the people of Salt Lake county and the counties north moving south into middle and southern Utah....	6,000,000
Loss sustained by crickets, locusts, and grasshoppers............................	2,500,000
Unsuccessful early experiments in making iron, sugar, paper, nails, leather, cotton-raising, mining, etc...................	6,000,000
Cost of defence against anti-polygamy legislation believed to be unconstitutional.....	3,000,000
Heavy freight rates from the Missouri river and the Pacific coast before the railroads	8,000,000
Cost of establishing the Overland Mail and Express Company, purchase of Fort Bridger, and establishment of Fort Supply, abandoned and afterward absorbed by the army of 1857...................	2,000,000
Protecting overland travel, succoring and feeding California, Oregon, and other emigrants...........................	1,500,000
Cost of colonizing Carson and Green River counties, abandoned because of the army of 1857........................	2,000,000
Cost of establishing colonies on Salmon	

river, in Lower California, and the sugar plantation near Honolulu..............	$1,500,000
Cost of local telegraph aad railroad lines....	3,000,000
Cost of obtaining fuel, and building and fencing materials, from the rugged mountains and canyons many miles away	10,000,000
Cost of making settlements on the Muddy, Call's Landing, Florence, Sunset, and other localities, afterwards abandoned because of adverse conditions subsequently developed............................	1,000,000
Losses by fire ($20,000 per annum).........	800,000
Taxes.....................................	8,000,000
Miscellaneous expenditures................	12,000,000
	$562,900,000
Less the personal property brought into Utah by immigrants, such as cattle, wagons, cash, etc.............................	20,000,000
	$542,900,000

In his note transmitting these figures Mr. Musser writes: "The inclosed has been submitted to the inspection of Presidents Woodruff, Cannon, and Smith, and Bishops Preston, Burton, and Winder, as well as to others conversant with such matters. All agree that the estimates are as fair as they can be given." And he adds, with a reverence characteristic of his people: "While much of our prosperity is due to industrious, temperate, and frugal habits of life, yet we never lose sight of the overruling hand of the Almighty in all these results, and to Him be given praise and thanksgiving without stint."

In a private letter accompanying these statistics Historian Musser directed attention to the fact that upon this showing each Mormon farmer enjoyed an average

THE CONQUEST OF ARID AMERICA

income of four hundred and eighty-two dollars *above* the cost of living for each of the more than forty years which the statement covers. This is a considerably higher return than the *gross* amount averaged by wage earners in the United States.

While in many particulars this imposing statement of results may be open to criticism, there can be no doubt that it was prepared with conscientious care. It is presented here for what it may be worth. To the writer it seems to confirm the impression of a vast material achievement which comes to any person upon visiting Utah and looking about him. For the present purpose the precise statistical facts are of less consequence than the economic principles which have produced what everybody acknowledges to be a wonderful result. These principles may be briefly summarized as follows:

GENERAL LAND OWNERSHIP, LIMITED TO THE AMOUNT WHICH FAMILIES AND INDIVIDUALS COULD APPLY TO A USEFUL PURPOSE.

SELF-SUFFICIENCY IN AGRICULTURE, AIMING AT THE COMPLETE ECONOMIC INDEPENDENCE OF THE PEOPLE, INDIVIDUALLY AND COLLECTIVELY.

THE PUBLIC OWNERSHIP OF PUBLIC UTILITIES, SUCH AS WATER SUPPLY FOR IRRIGATION AND DOMESTIC USES.

THE CO-OPERATIVE, OR ASSOCIATIVE, OWNERSHIP AND ADMINISTRATION OF STORES, FACTORIES, AND BANKS, THROUGH THE MEDIUM OF THE JOINT-STOCK COMPANY.

These are the underlying principles of the Mormon commonwealth. They are vindicated by the successful experience of the last half century. Nowhere else do so large a percentage of the people own their homes free from incumbrance. Nowhere else has labor received so

THE MORMON COMMONWEALTH

fair a share of what it has created. Nowhere else has the common prosperity been reared upon firmer foundations. Nowhere else are institutions more firmly buttressed or better capable of resisting violent economic revolutions. The thunder-cloud which passed over the land in 1893, leaving a path of commercial ruin from the Atlantic to the Pacific, was powerless to close the door of a single Mormon store, factory, or bank. Strong in prosperity, the co-operative industrial and commercial system stood immovable in the hour of wide-spread disaster. The solvency of these industries is scarcely more striking than the solvency of the farmers from whom they draw their strength. No other Governor, either in the West or in the East, is able to say what the Honorable Heber M. Wells said in assuming the chief magistracy of the new State in January, 1896. "We have in Utah," said the young Governor, "19,816 farms, and 17,684 of them are absolutely free of incumbrance." A higher percentage in school attendance and a lower percentage of illiterates than even in the State of Massachusetts, is another of Utah's proud records.

So far we have been dealing with facts that are beyond dispute. No one can deny that the Mormon industrial and commercial system is correctly described in the foregoing pages, nor that that system has made the people remarkably prosperous in an economic sense. But for the purposes of this book it is highly essential to determine just what weight should be given to the Mormon experience as a guide for future colonization effort in the arid West, and to what degree the Utah system is founded upon correct principles of industrial and social economy.

THE CONQUEST OF ARID AMERICA

The problem can be summed up in two questions which have doubtless already occurred to the reader: Was the Utah experience possible without Brigham Young? Was Brigham Young possible without the Church?

The first of these questions may be answered unhesitatingly in the negative. Without a Brigham Young there could have been no such record of achievement in the deserts of Utah. He was the brains and the soul of the enterprise. He planned with extraordinary sagacity and wrought with tremendous vigor. Leave out that brain and soul—that sagacity and vigor—and we can conceive of no emigration from Nauvoo; of no successful march over plain and mountain; of no triumph over the almost insuperable difficulties which intervened between the arrival of the people in Salt Lake Valley in 1847 and the firmly established community of fifty years later. But what of that? The concession of the indispensable fact of Brigham Young amounts only to the concession, equally applicable to all human undertakings of magnitude, that leadership is absolutely essential.

This brings us to the other and more complicated question: Was Brigham Young possible without the Church? First let us see what manner of man he was.

Born in Vermont, of good native stock, he had the characteristics of the place and the race in a pre-eminent degree. He was shrewd and thrifty, far-seeing and intensely practical. He was of coarse fibre, deficient in the finer feelings, and devoid of all imagination of the poetic kind. Of his innumerable sermons and speeches nothing survives save an occasional homely maxim, such as, "Plough deep and plant alfalfa." Like all his sayings and all his works, this marks the mind and method of the materialist rather than of the idealist. Whatever else he really thought of polygamy, he at least believed it a capital method of increasing the population of a new country, and that happened to be the particular work upon which his effort and ambition were engaged.

A leader of men? Most emphatically, but of the grim and masterful sort—a Cromwell rather than a Lincoln. While no orator, he had strong persuasive powers. These were supported by splendid enthusiasm and optimism. He could set men at work with the conviction in their minds that success was certain, failure impossible.

This man was successful in what he undertook to do. He did not originate Mormonism. He added nothing to its creed or its literature, though he added much to its power. But finding the Mormons a despised and hunted people, he set himself the task of extricating them from intolerable surroundings, of leading them a thousand miles across an almost unexplored country, and of founding, in the midst of untried conditions, a commonwealth where they could rear their homes and temples and wax great and strong. Who can doubt that if he had undertaken to build a transcontinental railroad, like Ames and Huntington; to found a pork-packing business, like Armour; or to lead an army, like Grant, he would have commanded success? He had all the elements of a successful man in any of the greater walks of life where pluck and brains, determination and vast ambition, are the requisite qualities. If he was a religious fanatic, there never was another of his composition.

THE CONQUEST OF ARID AMERICA

Poet or orator he could not have been; seer, revelator, and ecclesiastic he was not, save to the superstitious vision of his blind followers; but great, resourceful, and of commanding personality he was—a captain of industry, an organizer of prosperity; and the Utah of to-day is his undeniable claim to fame and his imperishable monument.

So much for the man. What of the Church? It was unquestionably the instrument used in the settlement of Utah. It is being used to-day as an instrument in settling portions of Canada, Mexico, and other localities. Regarded simply as a Church, it is successful numerically and financially. It is one of the few creeds where secular and religious affairs are brought into the closest association, and, for this reason, it is generally believed that church solidarity is the true explanation of the economic prosperity of the Mormons. This conclusion rests upon the theory that the Church sustains the industrial system. The writer emphatically dissents from this notion, and confidently asserts that precisely the reverse is the truth—that the industrial system sustains the Church.

The principles upon which the Mormon industrial and social structure was reared have been carefully presented in this chapter. These principles have worked successfully for fifty years. To determine the part which they had in the actual result, let us ask ourselves this question: Suppose the plans initiated by Brigham Young had failed to give his followers the security of a home and the certainty of a living; that their co-operative industry had produced losses rather than profits; that their village system had brought social discontent instead of

THE MORMON COMMONWEALTH

satisfaction—what then ? Is it conceivable that religious fanaticism could have held them together and lent such an impulse to their growth that to-day, over a quarter of a century after the death of Brigham Young, they should be growing faster than ever before, maintaining more missionaries and building more colonies in various parts of the world ? Surely economic fallacy never produced such striking results as these in any other instance known to history.

It would perhaps be a tenable position to say that in Utah a sound economic system, working in conjunction with religious enthusiasm, produced the result now known of all men ; but that would be very nearly equivalent to saying that the only way to solve the problem of reclamation and settlement in the arid regions is to turn the task over to the Mormon Church and to advise all who crave homes to join that organization. The writer believes that the attraction of Mormonism has consisted mostly in what it offered to the home-seeker, and that the secret of its cohesion is the prosperity that has resulted from its industrial system rather than the occult power of its creed.

Polygamy has so stirred the Christian world that no man may speak in praise of any of the Mormon institutions except at the risk of being misunderstood, or possibly regarded as an apologist for what the nation has condemned as a crime against womanhood. On the other hand, no candid mind can study the problem which confronts the American people—the problem of opening the door to the masses of our citizenship upon the unused natural resources of the nation—without realizing that Brigham Young and the State he founded furnish

THE CONQUEST OF ARID AMERICA

stronger and clearer light for the future of domestic colonization than any other experience that can possibly be discovered. It is in the earnest conviction that it is a high public service to show the virtues of the Mormon industrial system that this chapter is written.

CHAPTER II

THE GREELEY COLONY OF COLORADO

THE Greeley Colony of Colorado sprang belated from the seed of Fourierism sown broadcast in the forties. In all our social history there is no more interesting page than that which records the rise, progress, and temporary defeat of the doctrine of association. Fraught with the noblest aspirations, and welcomed and championed by the most brilliant minds, it disappointed, in actual practice, the high hopes of its friends. François Marie Charles Fourier devoted his life to elaborating his scheme of Socialism, and died a few years before the seed of his thought was wafted across the Atlantic to take sudden root in our soil.

The American impulse of Fourierism arose from the miseries of the hard winter of 1838. The doctrine had been imported by Albert Brisbane, a young gentleman of wealth and leisure who had studied the works of the French philosopher in Paris and returned to this country warm with these new hopes for humanity. Availing himself of the opportunity offered by the universal discontent, he plunged boldly into the agitation and attracted a remarkable degree of attention. Horace Greeley, then in the morning of his fame, espoused the new cause, at first cautiously, then with characteristic

THE CONQUEST OF ARID AMERICA

energy and daring. The period of agitation covered the years between 1840 and 1847. The men of thought soon won the confidence of the men of action, and a large number of associations for the purpose of bringing Fourierism to the practical test were formed in various States. In May, 1843, Mr. Greeley wrote in the *Tribune:* " The doctrine of association is spreading throughout the country with a rapidity which we did not anticipate, and of which we had but little hope. We receive papers from nearly all parts of the northern and western States, and some from the South, containing articles upon association, in which general views and outlines of the system are given. Efforts are making in various parts of this State, in Vermont, in Pennsylvania, Indiana, and Illinois, to establish associations, which will probably be successful in the course of the present year."

There was not much difficulty in obtaining recruits for these undertakings, and the experiment was entered upon with great enthusiasm. With a single exception, it ended in failure. The most famous of these colonies was Brook Farm, at West Roxbury, nine miles from Boston. Rev. George Ripley was the head of the enterprise. With him were associated, either as actual colonists or active sympathizers and supporters, Nathaniel Hawthorne, Ralph Waldo Emerson, Henry D. Thoreau, James Freeman Clarke, William Ellery Channing, Bronson Alcott, George Bancroft, Charles A. Dana, Margaret Fuller, and many others whose names rank high in the annals of American literature. Never before, and never afterwards, was such a galaxy of brains assembled in a single colony. Most of them were then in young manhood, with their fame all before them. But the historian

THE GREELEY COLONY OF COLORADO

of the enterprise sadly relates that, at the end of their first year they found they had a surplus of philosophers and a dearth of men who could hoe potatoes. And New England has been smiling about Brook Farm ever since. The end of Fourierism in the United States was the joint debate between Horace Greeley and Henry J. Raymond in their respective newspapers, the *Tribune* and the *Courier*, of New York.

In the minds of the devoted constituency of the New York *Tribune*, the idea of colony-planting as a means of improving the lot of average humanity had taken deep root, so that twenty-five years after Fourier's dream had ceased to flourish as a social experiment, a colony representing its hopes, if not its methods, could gain supporters.

The new venture was initiated by Nathan Cook Meeker, who had succeeded Solon Robinson as agricultural editor of the New York *Tribune* at the close of the war. In 1844 Mr. Meeker had been an active participant in the Trumbull Phalanx at Warron, Ohio. This had expired of ague, poverty, and dissension, after a fitful career of about three years. "If the place had been healthy," Mr. Meeker said afterwards, "we should have held out longer, and the idle and improvident would have got more out of the industrious and patient; but I have no reason to suppose that we should not have finally exploded, either in some fight, or at least in disgust." From this experience he emerged disappointed and destitute, but with valuable lessons for the future and unshaken faith in the utility of colonization effort. The knowledge thus dearly bought he was destined to apply, many years later, in a useful career as one of the founders of a State.

THE CONQUEST OF ARID AMERICA

In the fall of 1869 Mr. Meeker had returned from a trip to the Far West, the object of which was to describe the Mormon industrial system in a series of letters to the *Tribune*. Encountering a snow blockade at Cheyenne, which compelled him to postpone his visit to Utah, he had gone to Colorado instead. It was at the time when the Kansas Pacific Railroad was pushing across the plains to the budding village of Denver, transforming the wagon-trail into a highway of civilization. Everywhere Mr. Meeker beheld the dawn of a new industrial life in the midst of a wilderness. He was charmed with the climate and scenery, and impressed with the material wealth of the country's undeveloped resources. The old enthusiasm for colony-making filled his imagination. Wearied with a life struggle to remodel old social structures, he longed to avail himself of this opportunity to build on new foundations.

These hopes he communicated to his friend, John Russell Young, who agreed to bring the matter to the attention of Horace Greeley. This he did at a dinner held at Delmonico's in December, 1869. Mr. Greeley was instantly interested, and beckoned Mr. Meeker to join him at the table. "I understand you have a notion to start a colony to go to Colorado," said the editor. "Well," he continued, "I wish you would take hold of it, for I think it will be a great success, and if I could, I would go myself." Thus assured of powerful backing, Mr. Meeker at once proceeded to form his plans.

The prospectus of the new colony was drawn up by Mr. Meeker, but carefully weighed and revised by Mr. Greeley. A quarter of a century had elapsed since these men had been engaged—the one as active participant, the other as the most conspicuous American champion—in the Fourier scheme of association. It is interesting to observe just how much of the old plan survived in the new colony prospectus, when the thought of these leaders had been mellowed and broadened by many more years of life and experience.

In the Fourier communities the people had lived together under one roof, in the hope of effecting large household economies. There had been common ownership of land, and an attempt at equal division of labor. The unit of the community was the whole; the only individual, the public.

In forming the plan of the new colony the lessons of experience were not forgotten. There was but a single suggestion of the "phalanstery," or common household of Fourier days, and that was advanced in timid terms. "It seems to me," Mr. Meeker wrote, "that a laundry and bakery might be established, and the washing and baking done for all the community; but other household work should be done by the families." It was provided that the unit of society should be the family, living under its own roof; that farms and homes should be owned independently; that individuals should plan their own labor, and rise or fall by their industry and thrift, or lack of them. The new ideal was that of an organized community which should give the people the benefit of association without hampering individual enterprise and ability. It furnished a means of settlement essentially different from that under which the Middle West had been developed.

Land was to be purchased on a large scale with a common fund. This cheapened its cost, and gave the col-

THE CONQUEST OF ARID AMERICA

onists an important measure of control in its sub-division and development. The settlement was to be made almost wholly in a village, the land being divided into blocks of ten acres, and the blocks into eight lots for building purposes. It was proposed to apportion each family "from forty to eighty, even one hundred and sixty acres," adjoining the village. Northampton, in Massachusetts, and several other New England towns and villages, had been settled in this manner. A feature of much interest was the proposal to have the residence and business lots sold for the benefit of the colony's treasury, the capital so obtained to be appropriated for public improvements, such as building a church, a town-hall, and a school-house, and establishing a public library. This plan marked an important departure in town-making. Town sites, as a rule, especially where the community promises a rapid growth, are treated as opportunities for private speculation. The boom comes, and everybody prospers; the boom goes, and a few schemers have managed to acquire nearly all the cash capital. Under the new plan, as the prospectus pointed out, "the increased value of real estate will be for the benefit of all the people." They would receive these benefits, too, in the best form, as in the shape of permanent improvements essential to their social and intellectual well-being, and of capital available for industrial purposes.

Other advantages of settling in a village were presented as follows: "Easy access to schools and public places, meetings, lectures, and the like. In planting, in fruit-growing, and improving homes generally, the skill and experience of a few will be common to all, and much greater progress can be made than where each lives isolated. Refined society and all the advantages of an old country will be secured in a few years; while, on the contrary, where settlements are made by old methods people are obliged to wait twenty, forty, or more years."

This prospectus was published in the New York *Tribune* of December 14, 1869, with a hearty editorial indorsement. Spite of radical departures in the matter of private landholding and individual industry, the vital spirit of Fourierism lived and breathed through the cautious lines of the announcement. There was still the high ideal of social and civic life, of industrial independence, of a scheme of labor which should give to the laborer an equitable share of what he produced. There was still the plan of co-operation in achieving things for the common benefit. There was still the craving for a society composed of sober, temperate, industrious people. The common household had been discarded for the family home and hearth-stone, but for the barbarism and isolation of life on great farms there had been substituted the association of homes in the village centre, with the best social and intellectual opportunities. Behind the new plan, as behind the old, stood the patient energy and faith of Meeker and the glorious optimism of Greeley.

The announcement had met with a prompt and enthusiastic response at the hands of several hundred people, who had organized the Union Colony of Colorado at a meeting held at the Cooper Institute in New York, where Horace Greeley had presided. A committee had selected twelve thousand acres of railroad and government land in the valley of the Cache la Poudre, twenty miles northwest of Denver, on the line of railway then building to Cheyenne. The pioneers of the colony were thus able to begin settlement in the spring of 1870, and to bring to the test of actual experience the social and industrial plans set forth in the prospectus. A party of eastern people, most of whom came from cities, they entered cheerfully upon the task of adjusting a high ideal to the untried conditions of a country which had previously known only the Indian, the hunter, and the cowboy. Their experience for the next twenty years has a larger significance than merely local history, since the community is one of the landmarks in western life.

Mr. Meeker having refused the use of his own name, the new town was christened "Greeley," and this name was popularly applied to the colony also, in spite of its incorporated title. The first severe test of the co-operative principle, which had been relied upon for the larger enterprises, arose in connection with the building of canals. There had been no misconception as to the need of irrigation, but it was supposed that the works could be quickly constructed and the new methods of agriculture readily learned. The original estimate of cost was twenty thousand dollars. The actual outlay before the works were completed reached four hundred and twelve thousand, or more than twenty times the estimate. For resources to meet this unexpected demand, the colony had only receipts from the sales of property and the subscriptions and labor of its members. The result was not reached without serious dissensions and some desertions, but the works were built, and the community survived with its co-operative principle intact. It is not to be believed that a private enterprise could have lived through a similar experience with the same slender financial resources, for it was the public spirit and pride which

THE GREELEY COLONY OF COLORADO

saved the day at this critical juncture. These increased as difficulties multiplied, and rose with the tide of outside criticism and abuse. The process welded the people together, and made them strong enough to meet successfully the obstacles which yet remained.

Having provided water for their lands, the settlers proceeded to create the irrigation industry of Colorado; for nothing worthy of the name existed on the scattered ranches of the sparsely settled Territory. The newcomers brought their intelligence to bear upon the problem of perfecting skilful methods of irrigation and cultivation, and of discovering the classes of crops best adapted to the soil and climate. This work quickly led them to realize another disappointment of serious import. They had dreamed of orchards and vineyards, and of homes set in the midst of beautiful flowers and delicate shrubbery. Experiment soon taught them that they had been deceived about the character of the country. The hopes which had been built upon the fruit industry failed utterly, and the colonists were compelled to fall back upon general farming. This involved somewhat larger farms, and rendered more difficult the realization of their social plans. Very likely it saved them from the evils of the single crop which has marred the prosperity of many agricultural districts. The diversified products of the soil yielded them a comfortable living. Since there was no hope of obtaining cash income from fruit, they sought another surplus crop, and found it in the potato, to which their soil proved to be peculiarly adapted. They made an exhaustive study of this culture, and at last produced in the "Greeley potato" one of the famous crops of the West. Its superiority readily commands the best place in the market, and there have been years when the crop has returned a million dollars to the potato districts of which the colony is the centre. The farmers invented a pool system which frequently enabled them to control the output, and so influence prices in their favor.

Events proved that the colonists were gainers by reason of the trials and disappointments which attended the establishment of their industrial life. Though the cost of their canals had so far outrun their expectations, they obtained their water supply much cheaper than did subsequent communities who patronized private companies. At Greeley the cost of a water-right for eighty acres was three hundred and fifty dollars. This made the user a proportionate owner of the works. Where canals were private, settlers paid twelve hundred dollars for precisely the same amount of water, while the works remained the property of a foreign corporation. The difference in the price of water under the two systems represented a very handsome dividend for those who had persisted in their allegiance to the co-operative principle. In the same way, the colonists profited from their struggle to realize the best agricultural methods. They won a reputation for their products which possessed actual commercial value, and they became the teachers of irrigation, furnishing practical examples to students of the subject and contributing largely to its literature. These results must be credited to the fact that the community was organized, and that the people acted with a common impulse.

Passing now from the industrial to the civic side of the colony life, we find that the high public spirit in which the community was conceived left its marks not less indelibly. In the original prospectus Mr. Meeker had

THE GREELEY COLONY OF COLORADO

plainly stated, "The persons with whom I would be willing to associate must be temperance men and ambitious to establish good society." This was no sounding phrase, for the founder and his fellow-colonists wrote their principles into the title deeds which transferred farm and village property from the company to individuals. These provided that if intoxicating liquor were ever manufactured or sold on the land, title should immediately revert to the colony. The provision was enforced with splendid intolerance. Those who were not in accord with its spirit had not been invited to come, nor were they made comfortable while they stayed. Their unbending attitude on this subject gave the men of Greeley the title of "Puritans," which was a unique distinction in the Far West, in that day of cowboys and border ruffians. The prohibition clause in the deeds was stoutly resisted by a small minority, and went from court to court, until it was finally vindicated by the supreme tribunal at Washington. The Greeley local sentiment has always upheld the principle, and insisted that it was responsible for the admittedly high character of the community. Like several of the colony's plans, it has been extensively imitated.

The government of the community was vested in executive officers, but was actually ruled by public opinion. This found expression in numerous town meetings held in Colony Hall, which was one of the earliest buildings erected. Here all the public affairs were discussed with perfect frankness to the last detail, and no public officer ventured to stray far from the conclusions there pronounced.

Not even the early hardships and disappointments were permitted to mar the social life of the colony. The

THE CONQUEST OF ARID AMERICA

people made the most of the opportunities offered by the association of homes in the village, and organized a variety of social and intellectual diversions. At an early period an irreverent newspaper writer remarked: "The town of Greeley is a delectable arena, for of the entire population three-fourths are members of clubs that are eternally in session. Day may sink into night, flowers may bloom and fade, and the seasons roll round with the year, but Greeley clubs are unchangeable." In one of the letters by which Mr. Meeker kept the readers of the New York *Tribune* informed of the progress of the community, he spoke of these " overflowing meetings," and said: "In all our experience we have never seen such institutions so well sustained; and if we wanted to show strangers the best that is to be seen of Greeley we would have them visit the Lyceum."

David Boyd, who was both a prominent actor in these scenes and the historian of the colony, writes of the same subject, and throws a suggestive side-light on a notable trait of western life when he says: "In coming to a country which offered so many new questions for solution and presented so many new aspects of life, the minds even of those past their prime experienced a sort of rejuvenation. Being nearly all strangers to one another, each was ambitious to begin his new record as well as possible, and so put the best foot foremost." Here is the explanation of much of the superior energy which marks the life of new communities, and here lies the hope of social progress through colonization. The individuality all but obliterated in the great city springs anew and develops into blossom and fruitage in the fresh soil of colonial life. Institutions which would be quite impracticable in old and crowded centres get a footing in new countries, where men may exert untrammeled energies, and move freely in that atmosphere of social equality which is certain to characterize new communities and likely to endure while they continue small.

In considering the net results of Greeley Colony, it is important to note first that it has been thoroughly successful. In this respect it presents a striking contrast to the Fourier experiment, from which it may be said to have descended. Each man prospered according to his merit, and what the community undertook to do by means of co-operation it accomplished. It cannot be said that the latter principle was applied extensively. The capital realized from the sale of property was so largely absorbed in the construction of canals as to leave little surplus for other industrial and commercial enterprises. If one-half of this capital had been available for stores, banks, and small industries, it is likely that much which was necessarily left to private initiative would have been undertaken by the colony. In that case we should find broader lessons in co-operative effort than we do now. It it is also important to note that the community owed its prosperity to its high ideal and uncompromising public spirit. There was here no common religious tie as in the early New England colonies; no shadow of persecution such as that which bound the Mormon pioneers together in an indissoluble brotherhood. The nearest approach to this influence was the prohibition sentiment, and this formed but a small part of the original plan. These colonists were earnest men and women who had gone forth to make homes where they could combine industrial independence with social equality and intellectual opportunity. They were grimly determined to accomplish what they had undertaken. This spirit, and this alone, kept them from going to pieces during the first five years, and laid the foundation for their permanent prosperity.

Both Colorado and the arid West owe much to the example of Greeley. It lent an impulse to the development of their civic character, and made a deep and lasting impression upon their agricultural industry. The influence of the community on its immediate surroundings is yet more plainly visible. Its success resulted in large irrigation developments and numerous settlements in Colorado, Wyoming, and western Nebraska. A community without a pauper or a millionaire, Greeley has yet had a surplus both of men and of capital to contribute to the making of new districts. The colony of to-day is a well-built town of comfortable homes and substantial business blocks, surrounded by well-cultivated farms connected by a comprehensive canal system, which is the property of the land-owners. Although in periods of general business depression it has felt the heavy hand of hard times, few communities in the world possess a better assurance of a comfortable living in the future, while none has better educational and social advantages.

Horace Greeley followed the colony's development with the closest interest, writing frequent letters of advice, and even finding time to pay a hurried visit. His last letter to Mr. Meeker, written six days before his death, was as follows:

"FRIEND MEEKER,—I presume you have already drawn on me for the one thousand dollars to buy land. If you have not, please do so at once. I have not much money, and probably never shall have, but I believe in Union Colony and you, and consider this a good investment for my children."

To N. C. Meeker Mr. Greeley's death was, indeed, calamitous. Depriving him of necessary income from newspaper sources, as well as of financial backing in the colony operations, it made it necessary for him to seek employment in the public service, and this was directly responsible for his death. He was massacred by the Indians while serving as agent on the White River reservation. His work for the colony had been entirely unselfish, and his name deserves high rank among the founders of western civilization.

CHAPTER III

THE EVOLUTION OF SOUTHERN CALIFORNIA

THE most valuable lessons in all the romantic history of California may be found in a trivial corner of the great commonwealth. Upon a clear day the eye may readily scan its entire length from the San Timoteo hills to the shining sea. Between its parallel mountain ranges the width of the district seems but two or three miles, though in reality it is from ten to twenty miles. Ignoring the nomenclature of local districts, this is the San Bernardino Valley. It is upon this narrow territory that to a great degree the fame of California climate and productions rests. Here institutions have been created in the last thirty years which are destined to exert a powerful influence upon the future of the West.

What Holland was to the life of Europe in the fourteenth, fifteenth, and sixteenth centuries, southern California is to the life of the Pacific coast at the end of the nineteenth century. The industrial impulse which the men of the Netherlands caught from their conquest of the sea, the men of the southern valleys caught from their conquest of the desert. "Curbing the ocean and overflowing rivers with their dikes," says one of the closest students of Dutch history, "they came to love the soil, their own creation, and to till it with patient, almost tender care." So

EVOLUTION OF SOUTHERN CALIFORNIA

they became the fathers of scientific farming in Europe. They wrought a marvellous revolution in the methods of cultivating the soil. "When Catherine of Aragon wished for a salad she was compelled to send for it across the Channel by a special messenger." The civilization founded upon this wonderful agriculture maintained its high character through the whole range of their economic life. The habits of skilful industry which grew from the intensely cultivated soil conferred the same prosperity when adapted to the workshop and the store. The thread of co-operation spun from their common labor on the dikes ran through the entire industrial fabric of the crowded little nation. The influence of neighborly association involved in the conditions of existence on farms of petty size colored and shaped their social life. As it was in Holland, so it is in southern California.

The men of the southern valleys made the small-farm unit supreme. With marvellous patience and intelligence they worked out the highest methods of watering and tilling the soil known to the world. Tempering their speculative instincts with love of home, they developed towns and surroundings of rare beauty and comfort, and made them centres of high social and intellectual life. To compare these conditions with those which prevail in the great wheat- and cattle-ranches of the North, where labor is mostly servile, and where beauty has never laid its hand upon the home or dooryard, is like comparing Holland to Paraguay. Although the South has by no means escaped the evils of the single crop, it has vindicated irrigation and the small farm, and the extraordinary social possibilities inherent in both.

THE CONQUEST OF ARID AMERICA

These are the valuable lessons which may be set against the failures and disappointments of the last two decades.

In the stormy and heroic days of the gold epoch, of the Bear Flag, of the American conquest, and of the vigilance committees, southern California played a small part. Its past is the dreamy memory of old mission days, of peaceful shepherds, of great haciendas, of a land dominated by Spanish folk and speech. The land was a desert of sage-brush and cactus, in which a few scattered mission gardens made charming oases. Along moist river-bottoms there were sometimes fields and gardens, though not of the highest type. On the uplands light crops of wheat and barley were occasionally harvested, if spring rains happened to be fairly generous. But it was, apparently, a country which offered nothing to the stranger save climate and scenery. To this barren place came irrigation and the Anglo-Saxon, bringing a new era in their train.

The evolution of southern California may be studied in the experience of two representative colonies. These are Anaheim and Riverside. Both were undertaken by comparatively poor men, and made important contributions to the permanent prosperity of the district in which they settled. The success which they achieved and the methods by which they accomplished it colored and shaped the larger institutions which grew from these pioneer plantings. Anaheim owes its historical importance to the fact that it was the mother colony, but it gains added interest as an example of the way in which a number of petty capitalists may combine their means in large enterprises. It is useful, too, as showing the

CALIFORNIA CONTRAST—PICKING FLOWERS AT PASADENA, WITH THE SNOW SEVEN FEET DEEP ON MOUNT WILSON

outcome of the settlement of city workingmen on agricultural lands. Riverside represents a higher degree of social conditions, and is especially important and interesting as an example of the influence exerted by an entirely new element of population upon a country which had been neither developed nor appreciated by its natives and early settlers. A brief glance at the beginnings of these two communities is essential to any just comprehension of the condition and tendencies of the southern California of to-day.

Anaheim was projected nearly fifty years ago by Germans from San Francisco. They were all mechanics and small tradesmen, and each was possessed of a modest amount of savings. It was proposed that this capital should be united in a common fund and used for the purchase and improvement of a large tract of land. For this purpose a colony association was formed, the members paying one hundred dollars each and agreeing to make further contributions in monthly instalments. A committee was sent out to discover a good location and contract for its purchase. A body of land near the Santa Ana river, twenty-five miles southeast of Los Angeles, was chosen. A part of the colony was then detailed to build an irrigation canal, divide the land into twenty-acre farms, with a central village, and plant the whole tract in orchards and vineyards. In the mean time the main body of the association remained in San Francisco, earning money and sustaining the work in the field. When the colony had thus been completely prepared for occupancy, the settlers came with their families, building their houses in the village and assigning the farms to individuals by drawing lots. In order to make this di-

THE CONQUEST OF ARID AMERICA

vision equitable, those who obtained the choicest property paid a premium, which was divided among those to whom the poorer places had fallen. Most of the colonists devoted themselves exclusively to agriculture, but enough opened small shops and worked at their trades as blacksmiths, carpenters, painters, shoemakers, and tailors, to meet the needs of the community. With the division of the land the association settled its accounts, and only the irrigation canal remained public property. Co-operation had served an excellent purpose, however, in enabling the people to obtain their land at first cost, and to have it improved skilfully and economically in advance of their coming.

Beyond the hope of dwelling beneath their own roofs and working for themselves, the founders of Anaheim had brought no special ideal to the southern valley. They were people of common tastes, well content with simple prosperity and comfort. The community was thoroughly successful. It is also possible to record an almost uniform story of individual ease of life for the settlers. While a few became discouraged and sold out to their neighbors, much the greater number remained and became comfortably well off, while a few rose to wealth. They had come to the colony from the employments of city life, yet readily adapted themselves to the work of tilling the soil of their small farms. But the true importance of Anaheim was seen in the impulse which it gave to a new form of development in southern California. It had been a region of great ranches, where live-stock and grain held almost complete sway. Anaheim pointed the way to the subdivision of large estates and the intensive cultivation of the soil with the aid of irrigation. This demonstration was destined to work a revolution in the character of the people and country.

The Riverside Colony, perhaps the most widely celebrated of any of these communities, is a better example of the colonial life of California. In a truer sense than Anaheim, it is a product of irrigation, and it illustrates more fully than the mother colony the social possibilities inherent in this form of agriculture. Its history reveals a curious struggle between the forces of co-operation and of private enterprise, in the course of which both lent much strength to the colony and exerted a marked influence upon its fortunes. Like most of the pioneer settlements, Riverside was the dream of comparatively poor men who sought, in the fresh opportunities of a new country, better conditions for themselves and their children. The enterprise originated with Judge North, of Knoxville, Tennessee. His prospectus was issued from that place in the spring of 1870, and evoked a large response from many different States. In this prospectus the founder did not undertake to outline a social organization with any detail.

"Appreciating the advantages of associative settlement," ran the circular, "we aim to secure at least one hundred good families who can invest one thousand dollars each in the purchase of land; while at the same time we invite all good, industrious people to join us who can, by investing a smaller amount, contribute in any degree to the general prosperity." The advantage of co-operative over individual settlement was rather forcibly expressed: "Experience in the West has demonstrated that one hundred dollars invested in a colony is worth one thousand dollars invested in an isolated locality." That the projectors had formed a very decided opinion as to the most favorable location is evident in the following: "We do not expect to buy as much land for the same money in southern California as we could obtain in parts of Colorado or Wyoming; but we expect it to be worth more in proportion to cost than any other land we could purchase within the United States. It will cost something more to get to California than it would to reach the States this side of the mountains, but we are very confident that the superior advantages of soil and climate will compensate us many times over for this increased expense."

His circular had attracted the attention of a few men of considerable means, and with these Judge North set out for California to select the site of the undertaking. With the rare intuition which eastern men have frequently displayed in going to the West, the newcomers selected a location which seemed quite preposterous to the natives of the country. Planning the most ideal development which had thus far been attempted, they deliberately bought lands which had formerly been assessed at a valuation of seventy-five cents an acre. These lands then constituted a sheep pasture of inferior sort. They were similar to the stretch of desert which the transcontinental traveller sees in passing through Arizona. After the winter rains they bore a short-lived crop of wild flowers, but during most of the year they offered nothing more attractive than sage-brush and mesquite. The Mexican who owned them had not sufficient imagination to perceive how the new proprietors could realize a profit upon the modest sum of two dol-

EVOLUTION OF SOUTHERN CALIFORNIA

lars and a half an acre, for which he gladly sold them. But Judge North and his friends had two well-defined ideas in their brains. One was irrigation; the other, oranges. To the natives the first seemed impracticable, because of the expense; and the other ridiculous, because no one had ever raised oranges there upon a commercial scale.

The Santa Ana river rises in the Sierra Madre mountains, drawing its volume from a multitude of springs and canyon streams. It flows southwesterly for a distance of seventy miles, where it empties into the ocean. Riverside is about twenty miles from the source of the stream, and lies on the bluffs along its eastern bank. The conditions did not present such opportunities for the cheap and easy diversion of the waters as the Mormon pioneers found in Utah. In later years, as the demand for irrigation grew constantly larger and more insistent, it became necessary to resort to the very highest type of works for the distribution of water, and even the earliest canal required a cash outlay of fifty thousand dollars. Fortunately the capital was available, and thus the work of development went forward without faltering. The original canal was completed in the spring of 1871.

The enterprise had resolved itself into a private stock company, owning both the land and the water. The land was now sold to the colonists for twenty-five dollars an acre. This included the right to purchase a certain amount of water, for which there was an extra charge in the form of an annual rental. At the beginning this amounted to about one dollar an acre, but it rose with the demand for water, and the need of costly improvements in the system, until it reached an annual charge of ten dollars an acre.

In the experience of Riverside we may see the commercial romance of irrigation in its most striking form. The original sheep pasture, assessed at seventy-five cents an acre, sold readily at twenty-five dollars an acre when irrigation facilities had been supplied. While this represented a handsome profit to the original investors, it was extremely moderate compared with the returns which the second purchasers realized. A few years later the unimproved lands sold for prices ranging from three hundred to five hundred dollars per acre. The improved orange orchards, which had been evolved from the sheep pasture, were valued, and actually sold, at one thousand to two thousand dollars per acre. There have been years when the best of them earned a profit of fifty per cent. on the higher figure.

Riverside was destined to win its chief celebrity as the pioneer orange colony. Its founders had based their faith in the possibilities of this industry on what they had seen in the gardens of old missions.

They did not hesitate to plant their lands largely with citrus fruits in the face of many predictions of disaster. The new culture prospered from the start, but made severe demands upon the patience and intelligence of the settlers. During the same years in which the Greeley colonists were working out, by means of experiment and painful experience, the solution of agricultural problems for Colorado, the Riverside colonists were performing precisely the same service for southern California. The skill and the enterprise which the one people applied to potatoes, the other applied to oranges, with the same high results. The Riverside colonists not only exhausted their own sources of information on the subject of citrus culture, but induced the State Department at Washington to make its consuls in semi-tropical countries their agents. In this way they were enabled to learn all that foreign horticulturists knew about the business. They made constant progress in improving the standard of their fruit, their most marked triumph in this direction being the production of the Washington navel, or seedless, orange. Their orchards represented all the choicest varieties, which were cultivated with the highest skill. The original colony tract of two thousand acres has been gradually extended until it includes ten thousand. The shipment of oranges has risen to six thousand carloads annually, realizing about three million dollars.

The projector of Riverside had framed his prospectus on the lines of co-operative effort. We have seen that the enterprise speedily became private and speculative in character. This result was mostly due to the necessity of using large capital for the initial development, and to the fact that the colony included a group of individuals who possessed considerable means. Possibly the same result might have occurred in Utah if the Mormon pioneers had not enjoyed a fortunate equality in the matter of poverty. In Utah there was no capital except labor and brains, and these admitted of no other form of enterprise than pure co-operation.

The speculative instinct which took possession of Riverside and ran a mad race through southern California, accomplished much good, as well as much evil. And in the end the pioneer orange colony returned very closely to the original ideal of its founder. The principal irrigation system became in time the property of the people, and the water-rights were inseparably associated with the land. The orange-growers also found it necessary to seek refuge from the rapacity of the commission system in the adoption of co-operation for the sale of their product. Hence, in the two most vital features of their industry — the watering of their lands and the handling of their crops — Riverside is fully realizing to-day the hopes in which it was originally conceived. On the side of its social life it has never departed from its first ideal, and it is in this aspect that it may be studied to the best advantage.

The homes and avenues of this colony, which have been evolved from an inferior sheep pasture in less than a generation, are among the most beautiful in the world. In considering their widely celebrated charms, it should never be forgotten that these are the homes and surroundings of average people, and that they earn their living by tilling the soil. Making due allowance for climatic differences, there are equally beautiful residence districts in the suburbs of great eastern cities; but these belong to people who enjoy a degree of prosperity much above the average—to the small minority who are rich, or at least unusually well-to-do. They are not farmers, but business or professional men who have risen above the general level of society. At Riverside, on the other hand, at least ninety per cent. of the total population live in homes which front on beautiful boulevards, presenting to the passer an almost unbroken view of well-kept lawns, opulent flower-beds, and delicate shrubbery. Newspaper carriers canter through these streets delivering the local morning and evening dailies. Though this is a farming population, the homes are so close together that the people enjoy the convenience of free postal delivery. They fill their bath-tubs with water piped through the streets. They light their homes with electricity. In the centre of the colony they have fine stores, churches, hotels, and public halls. Their schools are of the highest standard, and are housed in buildings the beauty and convenience of which bespeak the good public taste. A well-patronized institution is the club-house and its reading-room. There is but a single saloon, and it is considered decidedly disreputable to frequent it.

The first result of the early colonies was to give a tremendous impetus to the settlement and development of southern California. The fruits of this new impulse are seen in the scores of charming communities which stretch eastward to the margin of the Colorado desert and southward to the border of Mexico. Redlands, Ontario, and Pomona are typical examples. The impressive city of Los Angeles, which grows alike in good times and in bad, is another product of the movement which traces back to the humble beginnings of these pioneer settlements established by a superior class of eastern emigrants. High land values and costly irrigation works have naturally resulted. But these are only the superficial evidences of economic forces which lie deeper, and which should be noted as the peculiar product of the colonial life of southern California.

The germ of Riverside, and of the civilization which it inaugurated in the San Bernardino Valley, is the small farm made possible by irrigation. This is alone responsible for the character of industrial and social institutions and of the people who sustain them. Where farms are very small—in Riverside they are from five to ten acres in size—they necessarily belong to the many. This means a class of small landed proprietors at the base of society. The condition is one which forbids the existence of a mass of servile labor like that which lives upon the cotton plantations of the South, and, to a greater or less extent, upon large farms everywhere, including the greater part of California itself. On a small farm the proprietary family does most of the work. Hence the main part of the population in such districts as Riverside is independent and self-employing.

The people of southern California are plainly moving along the line which leads to public ownership of public utilities and co-operative management of commercial affairs. But with them the movement is an economic growth rather than a political agitation. It is the logical outcome of their environment and necessities. A great body of producers and proprietors of the soil, they formerly stood between private irrigation systems, supplying the life-current of their fields, and private commission houses, furnishing the only outlet for their products. The condition was an intolerable one, since it made them utterly dependent upon agencies beyond their control. These instrumentalities the people are rapidly taking into their own hands, and it is inconceivable that they can ever again pass into private control. It is probable that California has seen almost the last of the attempts to establish the policy of private ownership of irrigation works, the most vital of all public utilities in arid regions. The system of co-operative fruit exchanges is carried forward by the same impulse. Already it handles more than half the enormous product. The producers have their own packing-houses, make cash advances to their members, and send their agents to represent them in distant markets.

It is pleasant to note that beautiful homes and high average prosperity have not spoiled the democratic simplicity of these communities. After the adjournment of the International Irrigation Congress at Los Angeles in 1893, its members enjoyed the hospitalities of many of the charming colonies in the neighborhood. In his remarks at a banquet tendered the party by the people of Santa Ana, Señor de Ybarrola, the representative of Mexico, paid a handsome compliment to the ladies who had waited upon the table. Afterwards one of the distinguished representatives of France remarked his surprise at hearing a public compliment to "the servants."

"What!" exclaimed Señor de Ybarrola, "did you think they were servants? Why, those were the leading ladies of Santa Ana."

"Do you mean to tell me," the French delegate demanded, in amazement, "that the leading ladies of Santa Ana put on aprons to serve strangers?"

"Certainly," the Mexican replied; "for in this country service is a title to respect."

The incident illustrates at once the hospitality and the equality which are characteristic of the social life of southern California.

CHAPTER IV

THE REVOLUTION ON THE PLAINS

THE semi-arid portion of the Great Plains constitutes a distinct division of the irrigation empire. Its history and its problems are peculiarly its own. During the last half century it has lived through three stirring and romantic epochs and entered upon a fourth. This last is one of absorbing human interest, and will doubtless shape the permanent civilization of the region.

When Francis Parkman and the Mormon pioneers traversed the country, late in the forties, it swarmed with herds of buffalo and tribes of hostile Indians. It was the era of savagery, broken only by the presence of a few frontier posts, which served as the occasional refuge of adventurers and hunters.

Almost miraculously the buffalo disappeared, and the red men retreated before the white wave which overflowed the western bank of the Mississippi and began gradually to people the eastern margin of the plains. Then the savagery of the desert suddenly gave way to the semi-barbarism of an epoch of cattle-kings and cowboys.

Just as the Indian and the trapper had surrendered to the cowboy and his herds, so the latter in their turn receded and largely disappeared before another element

THE REVOLUTION ON THE PLAINS

which now swiftly arose in the life of the Great Plains. The third era of American colonization, noted in a previous chapter, was yet at the stage of flood-tide. New railroads were pushing their iron highways westward across the prairie. Such *entrepôts* as Chicago, St. Paul, Omaha, and Kansas City were crowded with hopeful immigrants whose appetite for government land had been whetted by the stories of prosperity with which the newspapers teemed. Horace Greeley's famous injunction, "Go west, young man," still rang in the ears of ambitious youth and homeless middle-age. Land agents urged on the multitudes with a zeal born of the commissions on which it fed.

In the enthusiasm of the hour no one gave heed to the few croakers who hinted that there was somewhere a mysterious boundary-line beyond which all efforts at settlement must be disastrous. There was a theory that rainfall moved westward with population, and that the cultivation of the land wrought changes in climatic conditions. Under these circumstances it was not strange that the home-seeking hosts crossed the unknown boundary into the region of scant rainfall, and learned in hardship and bitterness the lessons which a more cautious and far-seeing government would have comprehended and taught to its children.

In the absence of such scientific determination of the conditions of the country, tens of thousands expended all their money and the most precious years of their lives in discovering what could *not* be done in the semi-arid region. The crushing and pathetic truth that nature had denied sufficient rainfall for the production of crops in a region where a multitude of people had made their

THE CONQUEST OF ARID AMERICA

homes dawned slowly upon the public mind, and the conclusion was stubbornly resisted. Between the acknowledgment of this fact and the beginning of practical efforts looking to the use of irrigation, there was a brief but exciting intermediate stage in which high hopes were built upon the possibility of precipitating rain by artificial means. An Australian genius suddenly appeared with a mysterious prescription warranted to assemble clouds in a clear sky and compel them to weep in the shape of copious showers. The end of this undertaking was the failure of the experiment and the suicide of the inventor. One of the railways discovered another wizard with another prescription, and hauled his special car over the entire length of its line, promising showers on regular schedule time. Even the Agricultural Department at Washington expended several thousand dollars in experiments in this direction. In this case, however, there was no mystery about the method adopted. It was the use of powerful explosives to be discharged at a high elevation. As nobody denied that heavy showers frequently followed great battles, and that it generally rained on the night of the Fourth of July, there were high hopes for the success of this undertaking, which occurred on an elaborate scale in Texas. Secretary Rusk described the preparations in detail, and summarized the outcome in the sententious remark: "The result was—a loud noise!" The theory exploded with the dynamite and disappeared from the minds of men with the last reverberation on the Texas prairies.

The mysterious line which divides the region of fairly reliable rainfall from the land of sunshine has been discovered at last and generally accepted. This, as stated

THE REVOLUTION ON THE PLAINS

before, is the ninety-seventh meridian west from Greenwich. It divides the United States almost exactly into halves, running through the middle of North Dakota, South Dakota, Nebraska, Kansas, Indian Territory, and Texas. The vast territory lying between this meridian and the foothills of the Rockies, bounded on the north by Canada and on the south by Mexico, is the semi-arid region of the Great Plains. Over all this vast district the tide of settlement had flowed and ebbed again, as we have seen. It now awaits the full development of the fourth epoch in its eventful and romantic history. The character and extent of this development is governed by the nature of the water supply, which differs materially in the several States.

The utility of irrigation on the plains was revealed in a curious way. In Finney county, near the western border of Kansas, thousands of acres were planted to wheat in the summer of 1878, and it seemed the sanest of projects to build a grist-mill to grind the crop. This was undertaken near the Arkansas river by enterprising merchants in the neighboring community of Garden City, but the new institution began and ended with a mill-race. Before the building and machinery were required, the wheat had surrendered to dry air and hot winds. Not an acre of the crop was harvested. And yet the blighted seed was destined to bear another and far more fateful crop and the forgotten mill-race on the banks of the Arkansas to grind a grist that would prove historic.

A few settlers remained to rake amid the ashes of their ruined hopes. Among them was a man who had learned the methods of irrigation while living in California and Colorado. It happened that his land adjoined the abandoned mill-race, and he readily obtained the right to turn the water upon a part of his farm. The result, though not surprising to the practised irrigator, was a revelation to his thoroughly disheartened neighbors. The soil which produced nothing in the previous summer responded to the new method of cultivation with enormous crops of all varieties of products. In quality they surpassed anything previously grown in that region. As these facts became known a new hope arose, like a star in the night, against the dark background of past discouragements. The Garden City "experiment" became the Mecca of students of irrigation throughout the wide region devastated by the drought. The ruined crop of the previous year and the useless mill-race gave birth to an influence which in fifteen years has assumed far-reaching proportions.

Kansas is the mother of irrigation on the plains. When the people heard of the miracle wrought by the waters of the abandoned mill-race their optimism instantly foretold a better civilization than they had dreamed of. Irrigation began here with canal-building in the valley of the Arkansas river. For a time the work was prosecuted with remarkable vigor. As early as 1890 over four hundred miles of large canals had been built, at a cost of nearly three million dollars. But the industry came suddenly face to face with an unexpected and almost fatal obstacle.

The Arkansas river rises in the mountains of Colorado and waters a broad and fertile valley before crossing the boundary into Kansas. In the upper State enterprise was busy with the diversion of its waters.

THE REVOLUTION ON THE PLAINS

In the absence of any regulation of interstate streams by national authority, the Colorado irrigators claimed the right to take the last drop of water for their own canals. This they proceeded to do during the growing season, leaving the canals of western Kansas as dry as its prairies. The investment of an English company in extensive works costing more than a million dollars was practically destroyed by this turn of affairs. There were many similar losses of less magnitude. It was at this stage that the lamented humorist "Bill Nye" remarked of some of the western rivers that "they are a mile wide and an inch thick—they have a large circulation, but very little influence."

When the Kansas irrigators found themselves deprived of their surface supplies they sought the underflow, and in the process of finding and utilizing it developed an entirely unique and very promising mode of irrigation.

The new experiment was first made at Garden City, within sight of the historic mill-race. It was found that in the Arkansas Valley water could be obtained by shallow wells ranging in depth from eight to twenty feet. This is raised by hundreds of wind-mills into hundreds of small reservoirs constructed at the highest point of each farm. The uniform eastward slope of the plains is seven feet to the mile. The indefatigable Kansas wind keeps the mills in active operation, and the reservoirs are always full of water, which is drawn off as it is required for purposes of irrigation. These small individual pumping-plants have certain advantages over the canal systems which prevail elsewhere. The irrigator has no entangling alliances with companies or co-operative associations, and is able to manage the water supply

THE CONQUEST OF ARID AMERICA

without deferring to the convenience of others, or yielding obedience to rules and regulations essential to the orderly administration of systems which supply large numbers of consumers. The original cost of such a plant, exclusive of the farmer's own labor in constructing his reservoirs and ditches, is two hundred dollars, and the plant suffices for ten acres. The farmer thus pays twenty dollars per acre (about double the average price paid to canal systems in this region) for a perpetual guaranty of sufficient "rain" to produce bountiful crops; but to this cost must be added two dollars per acre as the annual price of maintaining the system.

Farming under these conditions is limited to small areas, and intensive methods of cultivation become imperative. The result has been the evolution of a multitude of five-, ten-, and twenty-acre farms, each surrounded by its tall fringe of protecting cottonwoods, which inclose grounds variously planted to orchard, field, and garden. Perhaps these methods present a closer parallel to European agriculture than anything else found in this country, while the numerous windmills suggest comparison with Holland. Nowhere are there sharper contrasts than that which is presented by these green and fruitful farms, gleaming like islands of verdure upon the brown bosom of the far-stretching plains, which have been seared by the hot breath of rainless winds.

The uses of the artificial reservoirs are not limited to irrigation; they are usually stocked with fish, which multiply with surprising rapidity and enable the farmer to include this item of home produce in his bill of fare every day in the year. These fish are very tame, and in

THE REVOLUTION ON THE PLAINS

some cases actually trained to respond to the ringing of the dinner-bell, coming in scurrying shoals to fight for crumbs of bread thrown upon the water. (This fish story is a true one.) The reservoirs also yield a profitable crop of ice in the winter. When we compare the hardships and bitterness of this locality but a few years since with the comfort and abundance which the infinitely smaller farms yield to-day, we behold anew the civilizing power of irrigation. The Starvation Belt has become a Land of Plenty.

The centre and inspiration of these developments is Garden City, capital of Finney county. What Greeley was to Colorado and Riverside to southern California, this little town has been to western Kansas. Perhaps no other small place on the plains suffered a more violent attack of "boom" than Garden City in the feverish times of the last decade. Certainly none has held with more tenacity to its confidence in the final outcome of the country or contributed more to the early vindication of its faith.

It is difficult to estimate the reasonable possibilities of windmill irrigation in Kansas. There are enthusiasts who insist that the industry will be extended to nearly every acre, uplands as well as valleys. There are pessimists who assert that the amount of land reclaimable by such means is relatively very small. Of this subject the conservative hydrographer of the United States Geological Survey, Mr. Frederick Haynes Newell, speaks as follows:

"The existence of the subsurface waters of the river valleys of western Kansas has long been known. Like every other natural resource, its importance, at one time little recognized, has been seized upon by the so-called "boomers" and exaggerated to the extent of creating distrust and depreciation. It is, however, one of the most important of the natural advantages of the State, and one upon which the foundations of prosperity must be carefully laid. By a thorough employment of the underground waters, with the best methods, much of the vacant land of the State will be utilized for agriculture, and the remainder can become a source of revenue, indirectly at least. Taking the Arkansas Valley as best illustrating these conditions, the general statement may be made that water can be had everywhere within the valley at moderate depths, and in quantities such as to be inexhaustible to ordinary pumping machinery if properly installed."

Referring to the very much larger territory lying outside of the river valleys, the same authority says:

"In the portions of western and central Kansas where wells cannot be obtained at moderate depth, it will probably be practicable to store considerable volumes of water by closing the outlets of natural depressions. Favorable localities, although somewhat rare, can be found in nearly every county, and by the proper construction of substantial earth-dams considerable volumes of water can be held for use upon the lower lands. In one instance at least water thus stored has been pumped for use upon an orchard, and the success attained in this way should induce others to try similar devices."

The drought of 1890 made Nebraska one of the important irrigation States of the West. Canals had been built on the North Platte river near the Wyoming boundary, several years earlier, but the irrigation industry had won no general recognition. Thousands of farmers were persisting in the delusive hope of rainfall farming, and public sentiment was distinctly opposed to those who sought to include Nebraska in the arid region.

All this was changed by the events of 1890. In that year crops were ruined by dry weather and hot winds throughout a large part of the State, and the people in the western counties generally acknowledged that it was useless to longer persist in the effort to cultivate the soil without artificial moisture. Strangely enough, they seemed to draw a new inspiration from their blighted fields. Irrigation conventions were held at many county seats. The study of water resources, of methods and laws essential to their utilization, became earnest and general. The popular agitation rapidly crystallized into a permanent and organized movement which has gathered strength with each passing year. Comprehensive laws were enacted by the legislature and the office of State Engineer created. Meanwhile, large amounts of private capital were invested, many canals constructed, and the despised western counties began to rise in public esteem.

It is now clearly apparent that the very lands which refused to yield a return for the industry of the first settlers will sustain the densest population in the future and give the most absolute assurance of permanent prosperity. Already the time has come when a State irrigation fair can be held in western Nebraska and make a striking exhibition of results, and when a commonwealth which ten years ago resented as a libel the intimation that its rainfall was deficient, can proudly claim to rank

THE CONQUEST OF ARID AMERICA

among the greatest of irrigation States. The transformation which has occurred in public opinion is no less striking than that of the agricultural industry itself.

The State is more fortunate than some of its neighbors in the character and extent of its water supplies. Over its western boundary the North Platte pours a perennial stream of considerable volume, which feeds a number of large canals. The surface flow of the South Platte is mostly absorbed in Colorado, but when the two forks are united in Lincoln county they make a river of respectable proportions, which flows through the heart of the State and furnishes water both from its surface flow and from its gravel bed. The Loup river further increases the irrigation facilities in the central counties. In the southwestern part of the State the Republican and its tributaries supply a number of quite extensive irrigation systems. Along the northwestern boundary the Niobrara, a noble stream, is beginning to be utilized.

The conformation of the land in western Nebraska also offers more favorable opportunities for the storage of flood waters than are found in most of the prairie States. The possibility of irrigation from wells by means of pumps driven by windmills and by steam and gasoline engines, are also being thoroughly tested, with hopeful results. The experts of the Geological Survey report that even away from the river valleys, where the depth to water is considerable, small farms can be irrigated by this means at most points. This conservative authority estimates that fully one million and a half of acres can be irrigated in western Nebraska. Local enthusiasts put the amount very much higher, but even the former figure represents a reclaimed area three times greater than

THE REVOLUTION ON THE PLAINS

that on which the wonderful agricultural industry of Utah has been developed. The Dakotas are comparatively well watered by surface streams, but they flow in deep channels, and the uniform slope of the land to the eastward is only about one foot to the mile. Under these conditions it is not practicable to divert the flow by gravity canals, though it is sometimes done with the aid of pumping machinery. But the Dakotas rejoice in the possession of great artesian basins and of some of the largest flowing wells in the world. Many of them are one thousand feet in depth, and some of them furnish the remarkable flow of four thousand gallons per minute. Over sixteen hundred artesian wells were reported in these two States as early as 1891, and the number has constantly increased. The irrigation sentiment has been well organized and has resulted in the provision of progressive legislation.

Texas was also a severe sufferer from drought throughout the western part of its vast territory. The greater portion of it is well watered by rivers, by large perennial springs, and by artesian wells second only to those of Dakota. Here the people have also responded with high public spirit to the appeals of the irrigation champions, and the new era in the industrial life of the State is well under way.

The actual amount of land that may be reclaimed and cultivated in the semi-arid region furnishes no measure of the value of irrigation to this vast district. By enabling thousands to engage in farming, irrigation has made it possible to use the surrounding plains as the pasture for great numbers of beef cattle. In many instances small herds are owned by the farmers themselves, but to a large extent their crops are bought by those whose sole business is cattle-raising. Thus all the resources of the region are brought into use, and a wonderful prosperity has followed as the logical result.

From Canada to Mexico the revolution on the Great Plains is now in full tide. It is the most dramatic page in the history of American irrigation. It has saved an enormous district from lapsing into a condition of semi-barbarism. It has not only made human life secure, but revolutionized the industrial and social economy of the locality.

To a considerable extent it has replaced the quarter-section with the small farm and the single crop with diversified cultivation. It has transformed the speculative instincts of the people into a spirit of sober industrialism. It has raised the standard of living and improved the character of homes. It has planted the rose-bush and the pansies where only the sunflower cast its shadows, and it has twined the ivy and the honeysuckle over doors which formerly knew not the touch of beauty. It has made neighbors and society where once there were loneliness and heart-hunger. It has broken the chains of hopeless mortgages and crowned industry with independence.

Part Third

UNDEVELOPED AMERICA

"Mighty as has been our past, our resources have just been touched upon, and there is wealth beyond the Mississippi which, in the not distant future, will astonish even the dwellers on the shores of Lake Michigan.

"From the time my eyes first rested on the great uncultivated plains which lie between the Mississippi and the Pacific Ocean, my wakening dreams have been filled with visions of the incalculable wealth which the touch of living water will bring to life from those voiceless deserts. There wealth only can produce wealth, and man, singly and alone, might as well try to subdue the Himalayas as to cope with these wastes; but the hand of *united and associated* man is already reaching forth to grasp the great results.

"The same power which wastes millions on the Mississippi can be utilized to make the desert blossom with the homes of men, for whom and for all of us the now blighted soil will bring forth the fruits of the Garden of Eden."—Hon. Thomas B. Reed, *in a speech at Pittsburg*, 1894.

CHAPTER I

THE EMPIRE STATE OF THE PACIFIC

CALIFORNIA is widely celebrated, but little known. Its unique climate and productions, and the dramatic incidents of its early history, have been deeply impressed upon the popular imagination wherever the name of the Republic is spoken. These circumstances have given it rank among the most famous of American States; yet its problems and its future are inscrutable enigmas to all who have not studied the subject at close range, and to many who have. The anomaly that one of the States most talked of should be one of the least understood is not difficult to explain.

In the first place, California is known not by what millions of people have seen, but by what millions have read. Europe is better known by contact to Americans than California. A prominent American orator recently "discovered" California, and filled the newspapers with the interesting and suggestive impressions it had made upon his mind. He had been to Europe twenty times, and to the Pacific coast once, which is once oftener than many other distinguished travellers of the eastern seaboard.

Still further, the Anglo-Saxon race is dealing with new conditions in California. Coming from dense forests, from a land of heavy rainfall, and from a temperate

THE CONQUEST OF ARID AMERICA

climate where winters are long and stern, it settled in treeless deserts, in a land of slight and peculiar rainfall, and under a sky that never knows the winter.

Finally, California is in its infancy, having recently celebrated its fifty-fifth birthday as an American commonwealth. Born in a paroxysm of speculation—one of the wildest the world has seen— it has outlived a trying experience of lesser economic epilepsy, and come to the threshold of its true career strengthened and purified by the extraordinary process. In less than half a century several far-reaching changes have swept through the industrial and social life of the State, swiftly altering the conditions of labor and of business. Even for those living in the midst of these events it has been difficult to read their significance and estimate their influence on the ultimate character of the place and people.

What wonder, then, that to the outside world California has meantime appeared like a jumble of gold, palms, and oranges, of gilded millionaires and hopeless paupers, of enviable farmers living luxuriously on small sections of paradise, and of servile alien laborers herded in stifling tenements? Such are the conflicting aspects of the Golden State to those who view it from afar. What are the facts?

The literature of California is prolific. Perhaps no other locality in the United States has been so often written about. In dealing with a place which presents so many strange and fascinating features it is easy for praise to become extravagance. This is now so well understood that it is commonly thought that the words "Californian" and "veracity" are seldom synonymous. But the truth is that visitors from abroad have contributed

EMPIRE STATE OF THE PACIFIC

rather more than Californians themselves to the popular impression of the State and its wonders. It is the fleeting tourist rather than the permanent resident who becomes the more reckless partisan of the charming climate, the majestic scenery, and the vast resources which, to his exhilarated imagination, seem certain to burst into their full potentiality in the immediate future.

Without doubt, the most influential books ever written about California were those of Mr. Charles Nordhoff. His *California: for Health, Pleasure, and Residence* (1873), and *Northern California* (1874), had a great vogue at the time of their publication, and for many years after. They are as fresh and readable to-day as when written, and it is easy to understand why they should have exercised so powerful an influence in making public opinion. Mr. Nordhoff should not be confounded with the superficial enthusiasts who study the country only from car-windows and the verandas of luxurious hotels. Addressing his books "to travellers and settlers," he evidently realized the grave responsibility of the undertaking, and made a conscientious effort to describe the situation faithfully and conservatively. To keen observation, and a clear, vivid, descriptive style, he added a shrewd common-sense, which enabled him to divine, with striking accuracy, several important economic facts which the residents themselves overlooked or ignored. He went thoroughly over and into the country, accepting no facts at second-hand which it was possible for him to verify by personal investigation.

Nevertheless, he wrote as a tourist-correspondent, and is first among those of that class who have given California the place it holds in the popular imagination.

THE CONQUEST OF ARID AMERICA

Looking back now to his studies and the deductions he drew from them, it is interesting to note how conditions have changed in thirty years, and to what extent his words of advice require revision before they can be offered to the settler of to-day.

When Mr. Nordhoff wrote his books cattle and cattlemen were just beginning sullenly to recede before the rising tide of agriculturists in the great San Joaquin Valley. He correctly foretold the first effects of the industrial revolution that would follow, predicting that the railroad and the public lands, and, later, the old Spanish grants, would be divided among farmers; that the cattle would be compelled to seek the mountains for free range, and would come into the valleys only to be fattened upon alfalfa and other crops. But he foresaw only the first effects of these changes, and the farmer who should proceed upon his advice to-day would certainly fail to prosper.

Mr. Nordhoff championed the cause of the small farmer against the great landowner, but his idea of a small farmer is widely different from the present significance of the term. He saw in the San Joaquin "cheap farms for millions." These were to be acquired, either from the railroad or the government, in tracts ranging from one hundred and sixty to six hundred and forty acres. This was what he meant by "small farms," and they were small, indeed, compared with the great ranches of thousands or tens of thousands of acres. But they were still of quite imperial dimensions compared with the unit of ten, twenty, or thirty acres which is now considered amply sufficient for the settler's needs.

While Mr. Nordhoff recognized the advantage of irrigation, he did not appreciate its actual importance, nor did he realize how largely it would increase the cost of land and how seriously it would influence the entire economic character of the country. He held out the hope of a prosperous living for families of small means who should settle upon farms of one hundred and sixty acres and upwards in the San Joaquin Valley, and depend chiefly upon crops that could be grown without irrigation. If "the millions" had accepted this advice in the past, or should do so to-day, nothing but disaster could result. Except in a few localities, prosperous agriculture in the San Joaquin Valley without irrigation is impossible. The character of the country is such that large and costly canal systems are required to bring any considerable portion of it under water. When these were built it was no longer possible to acquire cheap land, and the size of the practicable farm unit had been reduced to about one-tenth of the amount Mr. Nordhoff advised. These developments changed the situation completely.

The enthusiastic author was by no means blind to the possibilities of horticulture, nor did he fail to foresee that when this had been established it could be successfully pursued on much smaller areas. But here also his advice is now quite obsolete, and must be revised before it can again be offered to the public. He left the impression that oranges could be grown throughout southern California and the San Joaquin Valley. Later experience has eliminated the dream of orange orchards from a vast portion of these localities, but has demonstrated that the industry is practicable in some places where it was formerly supposed to be out of the question. While the orange-tree will grow and generally

THE CONQUEST OF ARID AMERICA

bear fruit throughout the lower valleys, the area in which it can successfully be cultivated for commercial purposes is rather severely restricted. To grow a few orange-trees within the shelter of the house, and to produce sufficient fruit for home purposes, is one thing; to grow thousands of acres of oranges fit for the market, and thus develop a genuine citrus district, is entirely different. There is a well-recognized thermal belt in the foothills of the Sierras, bordering the San Joaquin and Sacramento valleys, but the conditions of the country as a whole, with reference to this subject, have turned out to be very different from what they were supposed to be when Mr. Nordhoff wrote his books. In southern California his predictions in regard to orange culture have been largely realized, but even there it has been discovered that the field is limited.

The author was not unnaturally led into the error of saying that "the seasons are a little later in the North" than in the South. The contrary is the case, strange as it may seem, for it is the northern fruit districts which send the earliest products to market. This is true of both deciduous and citrus fruits. In the case of the latter the difference is very striking, as the northern oranges are ready for the Thanksgiving market, while comparatively little of the southern crop is available for Christmas purposes. Both the raisin and the prune industries were beginning to assume importance in 1873. Mr. Nordhoff quoted raisins at "two dollars per box of twenty-five pounds," and added: "I judge from the testimony of different persons that at seven cents per pound raisins will pay the farmer very well." To-day they are quite content to obtain three cents. He quoted prunes as bringing from twenty to twenty-two cents at wholesale at San Francisco, "and even as high as thirty cents for best quality." Prunes now bring from three to eight cents, and pay well at four and a half. Figs were then selling at from five to ten cents per pound, and the author thought they would be very profitable. The result has proved that while figs bear most prolific crops they are not profitable, as Californians have not yet been able to cure and pack them successfully. There are exceptions to the rule, but this is true as a general statement, and the fig is not a profitable article of commerce in California. In much the same way tobacco-culture failed and disappointed the hopes which had been built upon that industry.

These are instances of many particulars in which even the most painstaking of works on California require revision in the light of experience. So, too, the public opinion which they helped to make must be revised. Mr. Nordhoff described California as it looked and as it seemed to promise in 1873. While his methods were conscientious, his tone was one of intense enthusiasm. His vision extended as far as any one's could do at that time. The fact is that at that stage of its history California had not begun to develop its real and enduring economic traits as it has done during the past few years. It had recently emerged from an era of wild speculation. It stood upon the verge of another, in which railroads and agriculture, rather than gold, were to be the principal factors. It is from the calm sea-level of these quiet days that the State may best take its bearings. Thus the time is ripe for a new study of what in many respects is the most wonderful of American States.

THE CONQUEST OF ARID AMERICA

The great farmer of California is the successor of the gold-hunter. Both were speculators of the thoroughbred type; both looked with contempt upon the matter of making a living, and dreamed only of making a fortune. Of homes and institutions they were neither architects nor builders, for they sought only to take the wealth from the soil and spend it elsewhere. The miner leaves nothing to commemorate the place where he gathered gold save crumbling hovels and empty tin cans. The five-thousand-acre wheat-farmer leaves no monument beyond fields of repulsive stubble and the shanties of his "hoboes." These social forces belong to barbarism rather than to civilization.

Mr. Nordhoff clearly perceived these things, and not only urged the importance of smaller farms, but that farmers should be encouraged to diversify their products and become independent on their own places. But the conditions were yet too favorable for speculation. Wheat commanded more than one dollar per bushel. Of the new products, such as raisins, prunes, and oranges, the output was slight, and the prices consequently high. The result was inevitable. The owners of large farms sought to buy more land and increase the scale of their operations. The new settlers acquired as much land as they could, while the growing class of horticulturists planted their property exclusively to the few kinds of trees or vines which seemed most profitable at that time. Writing of this subject Mr. T. S. Van Dyke says: "The general principle upon which all farming was done, from the highest to the lowest, was very nearly this: Do nothing yourself that you can hire any one else to do, make no machinery at home, and raise nothing to eat that you can buy."

EMPIRE STATE OF THE PACIFIC

The rise of horticulture brought no material change in these conditions. As with the miner and wheat-farmer, so with the fruit-grower the aim was to get rich quickly, and the method speculation. Certain districts were devoted exclusively to prunes, others to wine grapes, others to raisins, and yet others to oranges. Fruit-land rose to almost fabulous prices, and was readily bought by those who had been taught to believe that they could realize profits ranging from one hundred to one thousand dollars per acre for certain crops. Exceptional instances justified this prediction, and everybody seemed to prefer to found expectations upon these instances rather than upon average returns. It is not difficult to understand why a man who counts upon an income of five to ten thousand dollars from ten acres, or double that amount from twenty acres, should turn his back upon common things, and devote his land exclusively to the crops which promise such gilded profits.

This was the general policy, and it conferred great prosperity upon some classes, particularly the Chinese and Italian market-gardeners, who raised food for the gentlemen-farmers to eat. There were years, however, when the fruit of trees and vines brought very large returns. Wherever the policy of single crops is pursued, whether it be wheat, corn, or cotton, raisins, prunes, or oranges, there are occasional years of well-nigh riotous prosperity. But such years are frequently more disastrous in their results than sober periods of depression. They feed the flame of speculation and raise false industrial ideals. Under the spell of such times, the people depart still further from the safe path of self-sufficient agriculture, buying more land to devote to the favorite

THE CONQUEST OF ARID AMERICA

crop, expanding their living expenses, and running into debt. When this spirit becomes the breath of industry no human laws can avert disaster.

A true industrial system is like a noble river fed by eternal snows: it never floods its banks with an excessive flow, and never sinks below its normal stage. It ebbs and flows with the regular tides of the great commercial ocean to which it is tributary, but alike at high water and at low, it bears the ships of men upon its tranquil bosom.

After a very intimate acquaintance with California horticulture, and with the army of producers who have engaged in it, Mr. Edward F. Adams, formerly manager of the State Fruit Exchange, wrote as follows :

"Unless certain reforms in the trade can be effected, there is danger that a large portion of the capital will be lost. The mortgage indebtedness is very serious; the general depression in values has temporarily wiped out the equities of the nominal owners; and while a partial recovery is doubtless to be expected in due time, it is not believed by the best informed that under present conditions of marketing, our orchards and vineyards can continue to maintain those who occupy them in their present standard of comfort. We are endeavoring by a general popular movement to remove the evils which oppress us."

Notwithstanding such warnings as this, and the sore experience on which they are based, there are real-estate interests which still advertise the fabulous profits of California fruit-culture, and there are many who believe them and proceed to organize their farms in the old way.

EMPIRE STATE OF THE PACIFIC

The evolutionary process of the last twenty years has wrought out some very valuable lessons for the future of California. It has demonstrated that irrigation is essential to the highest standard of civilization. The census of 1900 revealed the fact that nearly one-half the gain in rural population stood to the credit of eleven counties where irrigation prevailed. The counties which rely upon rainfall had about reached a stand-still or scored a loss. The people have always been divided on the question as to whether irrigation is necessary. Those who oppose urge that it breeds malaria and injures the quality of the fruit. Those who favor insist that it is essential to the most scientific agriculture, and to the maintenance of dense population. The last twenty years have answered the question forever. The answer consists of a comparison between the south and the north. The one was born of the irrigation canal; the other of the mining-camp and the wheat-ranch. The one is characterized by a high civilization; the other by a low one.

With a population, according to the census of 1900, of less than a million and a half, California has a territory nearly as large as that of France. It is inferior to France neither in climate, soil, natural resources, nor sea-coast, and its capacity for sustaining a dense population is fully as great as that of the European republic. The latter supports more than thirty-eight millions. If, then, the comparatively few inhabitants of the California of to-day are not equally prosperous, it is because they have failed to make the best use of their opportunities. With the same rate of increase in the next century as in that of the immediate past, the United States will contain in the year 2000 a population of over one thousand millions. Nothing is more certain than that California must receive its full share of these future millions. It seems hardly less certain that they will realize there the highest destiny of the race. But how?

Notwithstanding the supreme attractions of its rural life, more than seventy-three per cent. of California's total increase in the last decade covered by the national census settled in towns and cities. As a result, the urban life of this far, new State is as badly congested as that of the old communities of the East. But the possibilities of agriculture, of manufacture, and of mining are relatively untouched. Ultimate California remains to be fashioned from these undeveloped materials. The tendencies of future growth are revealed by the teaching of the past, and not less by its failures than by its successes—not less by the fury of old speculations than by the calm current of these saner times.

The future tides of population in the Golden State must first spend their energy upon the soil. It is the creation of a new and ampler civilization that is involved, and agriculture must be its foundation. But if those now engaged in cultivating the soil can scarcely maintain themselves, what hope is there for new recruits in the industry? The question is natural, but the answer is conclusive. There is no hope for them if they engage in speculation, but there is an absolute guaranty of a living and a competence, to be enjoyed under the most satisfying and ennobling social conditions, if they work upon sound industrial lines. These lines are clearly disclosed by the light of past experience.

Three classes of products should enter into the cal-

EMPIRE STATE OF THE PACIFIC

culations of the new settler in California: the things he consumes; the things California now imports from eastern States and foreign countries; the things which eastern communities consume, but can never hope to produce, and of which California possesses virtually a monopoly. In the first list is almost everything which would appear in an elaborate dinner menu, from the course of olives to the course of oranges, nuts, and raisins, and excluding only the coffee. This policy of self-sustenance has been ignored to a startling degree in the mad struggle for riches, but the coming millions of farmers can be sure of a luxurious living only by stooping to collect it from the soil.

In the second list are many of the commonest articles of consumption, which California might readily produce at home, but for which it sends millions of dollars abroad each year. The imports of pork and its products range as high as eight or ten millions each year. Condensed milk is not only a very important article of consumption in mining-camps and great ranches, but is largely shipped abroad for the Asiatic trade. It is brought across the continent from New Jersey. California also sends beyond its borders from twenty to twenty-five millions annually for the item of sugar, which should not only be produced in sufficient quantities to supply consumption, but for export as well. It is a curious fact that many of the finest fruit preserves sold in San Francisco bear French and Italian labels, and that the supply of canned sweet corn comes mostly from Maine. Essential oils made from the peelings of citrus fruits are also imported. It is not uncommon to find orange marmalade which has been prepared in Rochester,

THE CONQUEST OF ARID AMERICA

New York, the oranges having been shipped eastward, and the manufactured product westward, at the cost of two transcontinental freights. Imports are by no means confined to things which require capital and machinery for their manufacture. Chickens, turkeys, and eggs are largely brought from outside. A single commission-house in San Francisco imports five hundred thousand chickens every year. It is true that during the last few years notable progress has been made in sugar manufacture and in some other lines of production formerly altogether neglected, yet many thousands of new settlers can be profitably employed in feeding the growing home population.

Having made perfectly sure of his living, and disposed of his surplus for cash in the home market, the settler still has left a promising field in the list of things which nine-tenths of the American people consume but cannot produce. Among these products are oranges, lemons, and limes. Florida competition in this line has been temporarily destroyed, if not permanently injured. Mexico is, perhaps, a rising competitor; but there is little reason to fear that California cannot hold its own against all foreign producers. Even more promising is the olive-culture; for while the orange is an article of luxury, the olive must ultimately become here as elsewhere an important article of food. Californians are just beginning to pickle the ripe olives. The difference between a green olive and a ripe one is precisely the difference between a green and a ripe apple. In Spain the people subsist largely on olives—but not on green ones. All who have eaten the ripe fruit which is now being pickled in California will agree that it is conservative to say that when the American public become acquainted with this product, its consumption will be enormously increased. This will be true, because in its new form the olive is as nutritious as it is palatable, and the people will learn to depend upon it as an article of diet. In the production of deciduous fruits, such as peaches, apricots, cherries, and nectarines, California has much competition, and is to have much more in the future. There are irrigated valleys throughout the Pacific Northwest, the intermountain region, and the now undeveloped Southwest, which are beginning to produce marvellous fruits of this kind. The same is true of olives, almonds, and walnuts in a much more restricted way. The California wine industry is promising to-day, and the culture of grapes for this purpose profitable. Planters who depend for their entire income upon the cultivation of these export crops will necessarily suffer all the evils of speculative farming, but those who have founded their industry upon the plan of self-sufficiency will always have a surplus income from this third source, and in years of high prices it will be large. It is thus that the agricultural basis of California will be indefinitely broadened in order to sustain future millions.

Upon this foundation manufactures, mining, and an enlarged commerce will rest. The first cannot be long delayed. California will not permanently endure the enormous waste involved in shipping its wool and hides across the continent to Eastern mills, tanneries, and workshops, and in shipping back again the manufactured cloth and shoes. The factories must inevitably grow up near the raw material and the consumers. Expediency and the economy of nature alike demand it. This important part of California's civilization remains almost

THE CONQUEST OF ARID AMERICA

wholly to be developed. Its growth will open new avenues for employment and new outlets for the products of the soil.

The mining industry is also in its youth. To use a common phrase, but a true one, "the surface of the ground has only been scratched." Old methods have been outlived, and the conditions of the industry are changing in vital ways; but the work of taking gold and silver, copper, lead, and iron from the foot-hills and mountains of California has only been begun. The day of the individual miner, working with his pan in the gravel bed of the stream, is mostly passed. The conditions of hydraulic mining were materially altered by legislation because of the injury done by polluting the rivers and filling their channels; but quartz-mining is in a state of rapid development, and is destined to assume prodigious proportions. It will add untold millions to the wealth of the community, increasing the demand for labor and widening the markets of the farmer.

Nature has unquestionably provided the foundation of a marvellous industrial life in which millions of people will finally participate. To-day these resources are undeveloped. There is but one force that can awaken the sleeping potentialities into a manifold and fruitful life. That force is human labor. Looking down the years of the future, it is possible to predict, with the accuracy of mathematics, that human labor will coin from these vacant valleys and rugged mountain-sides billions upon billions of money. The wealth to be so created will build many beautiful homes, capitalize banks, factories, and railroads, and send great steamships across the Pacific to foreign shores. To whom shall these things be-

EMPIRE STATE OF THE PACIFIC

long when labor has made them from the materials which nature provided? Upon the answer to that question hang the destinies of California.

The seed of the California of the past was in the little group of feverish gold-hunters who camped by Sutter's mill in 1849. It bore the gaudy weed of speculation, with its bitter harvest of misfortune and discontent for the many, accentuated only by the superfluous riches which it gave to the few. The seed of the California of the future is in the irrigation canals owned and administered by small landed proprietors; in the fruit exchanges, which are supplanting the commission system and securing to the producer the rewards of his labor; in the co-operative creameries and canning factories which, in the face of deficient capital and unfair competition, are slowly fighting their way to the sure ground of abiding prosperity; in the multitudinous and uniformly successful manufacturing and mercantile associations which Mormon genius has planted in the valleys of Utah; in the banks, insurance companies, and loan and building societies which, all over the Union and all over the world, have vindicated the possibilities of associated man.

It is interesting to consider what portions of California will receive the bulk of the future population. The topography of the State is peculiar and readily comprehended. The coast region presents a frontage of over one thousand miles to the sea, and is narrowly hemmed in by mountain ranges which, in many places, come down to the shore itself. But in these mountains there are many picturesque and fertile valleys which have long been applied to agricultural purposes. The coast region

THE CONQUEST OF ARID AMERICA

has a climate of its own. It is the mildest type of the temperate zone, closely verging upon the semi-tropical, but not adapted to the growth of citrus fruits. Here the rainfall is heavier than elsewhere in California, and proximity to the sea gives rise to frequent fogs. In the southern extremity of this region, from Santa Barbara to San Diego, the climate becomes genuinely semi-tropical and fogs are less common. North of San Francisco the leading industries are lumbering, dairying, stock-raising, and general farming, with some mining. In a few favored valleys fruit-raising on small farms is successfully followed. South of San Franciso the lumber and mining interests are insignificant, and the country is mostly devoted to dairy, stock, and general farming.

A most notable exception to what has been said of the general condition of the coast region is the Santa Clara Valley, which contributes enormously to the exports of the State. In the beauty of its homes and orchards and the excellence of its horticultural methods, in the organization of its fruit exchanges, and the character of its urban life and civic institutions, the Santa Clara Valley is fully equal to the most ideal localities in California, not even excepting the famous orange districts near Los Angeles. There are numerous opportunities in counties farther south, notably in Monterey, San Luis Obispo, and Santa Barbara, to apply the same methods with similar results. But while the Santa Clara Valley represents the finest possibilities of the coast region, it also strikingly illustrates certain failings in the economic system of the State which have been dwelt upon in earlier pages. Land is almost exclusively devoted to fruit. Farmers buy their milk, butter, eggs, poultry, bacon, and fresh meats of others. They themselves produce none of the real necessaries of life, but only the luxuries. One reason for this is the lack of irrigation. They have taught themselves to believe that this is not only unnecessary, but would actually be injurious to the quality of their fruit. They are learning gradually, however, that this idea is erroneous — that skilful and proper irrigation is always beneficial, and that artificial moisture is imperatively necessary to diversified production; hence, to the highest business prosperity and best social conditions. When this lesson is learned by the coast region as a whole a new era will set in, and great numbers of colonists will come.

What is popularly known as southern California is a narrowly restricted district reaching eastward from Los Angeles for about one hundred miles and southward to San Diego. Like the coast region, its character is fixed, though on widely different lines. Its population is already comparatively dense, and its future growth will be measured by the water supply for irrigation. While it would seem as if the water resources had been fully utilized, the fact is that large quantities run to waste in seasons of flood, and that the cultivable area can be gradually extended by storage works and more economical methods of irrigation.

It is an impressive fact that the seven counties of the south received ninety per cent. of the increase of rural population between 1890 and 1900. This marvellous showing was chiefly due to the superior public spirit of the locality, and to the attractive institutions which grew out of it. Los Angeles itself is the throbbing heart of a region which, in many respects, has no equal

THE CONQUEST OF ARID AMERICA

in the world. The leading characteristics of this locality have been referred to in another chapter. But the very success which attended these methods in the past places limitations upon the country as a field for future expansion. Land values have risen high and the water supply has become almost as precious as gold. Health-seekers and the leisure class have been attracted in large numbers and occupy the field which would otherwise be open to home-makers of smaller means. A class of weathy people is a prominent feature of immigration in the southern valley. These opulent settlers plant orchards of oranges, lemons, and olives, just as their poorer neighbors do. It is reassuring to reflect, however, that they can accomplish little more with their abundant capital than humbler settlers may do with their united labor. The sun, the sky, the earth, and the waters will be as kind to one class as to the other. While it should not be inferred that none but the very rich can settle in the south, it is perfectly true that this charming district is not within the field of the largest future developments.

Where, then, is the field to accommodate the hosts who will come when the population of California begins to approximate that of France? It lies principally in four great and distinct bodies, which may be named, in the order of their importance, as follows: the Sacramento Valley, stretching north from the bay of San Francisco to the feet of snowy Shasta; the San Joaquin Valley, reaching south from the great bay to the place where the two mountain-ranges meet at the pass of Tehachapi; the intermountain valleys on the eastern slope of the Sierra, extending over the boundary into Nevada; and the Colorado Desert, in the extreme southeastern part of the state, on the borders of Mexico.

The first of these, the Valley of the Sacramento, presents one of the most remarkable opportunities for colonization to be found in the world. Although it represents rather less than one-half of the great interior valley of California, its length is equal to the distance from New York to Richmond, Virginia. Unlike other parts of the State, it is magnificently favored in its water supply. Government experts declare that ten million acres could be irrigated, and it is probably within bounds to say that a total of ten million people could be sustained, in town and country, when the resources of the region are brought under full development. It is rich in timber and minerals, while its climate favors the production of everything for which California is famous. Although hundreds of miles north of Los Angeles, it produces the earliest fruits, including oranges. It has navigable streams, capable of much improvement. The Sacramento River itself is valuable for commerce for a distance of nearly two hundred miles north of the bay of San Francisco.

But with all these remarkable advantages, the rural population increased but a beggarly thousand in the decade covered by the last census (1890-1900). The gain in town population during the same period was 9,240, making a net gain of a little over ten thousand for the Sacramento Valley as a whole.

What is the explanation of this stagnation in rural settlement? The country is held in large estates, principally devoted to the cultivation of grain, which is always

THE CONQUEST OF ARID AMERICA

a speculative industry, with alternating periods of prosperity and depression. Even horticulture is often conducted on a great scale in this region. The orchards and vineyards of the Stanford and Bidwell estates are striking examples of this tendency. The public spirit which gave the southern counties their splendid place in the life of the Pacific Coast is lacking in the north, though strongly represented by an aggressive and persistent minority whose influence is apparently gaining ground. The truth is that such public spirit is cultivated only with the greatest difficulty in a land of wheat fields and mining camps. It comes with irrigation, with the subdivision of land into thousands of small holdings, with a citizenship composed of a multitude of small proprietors.

Irrigation is by no means absolutely necessary in the Sacramento Valley. If it were, the story of its progress would be different. No one could then truthfully assert, as now, that this splendid district sustains less population on the soil than it did a quarter of a century since. While irrigation is not indispensable, it is essential to the best and highest results, especially in the line of small-farming. The rainless season usually extends from May until November. Without artificial moisture there can be no beautiful lawns, successive crops of vegetables and small fruits, or goodly yields of alfalfa. Citrus fruits cannot be profitably cultivated without it, and there is no fruit that is not improved, both in quality and quantity, by the proper application of water. This claim is often stoutly disputed, particularly by those wishing to sell land that is unprovided with irrigation facilities. But experience has taught that northern California may only hope to equal the southern part of the State by imitating its industrial methods, of which irrigation is first and foremost.

A striking example of what can be done upon these rich lands by means of irrigation and intensive cultivation is furnished by the experience of Samuel C. Cleek, who lived for many years on a single acre at the little town of Orland, in Glenn County. On this one acre he not only supported himself and wife in generous comfort, but averaged a cash saving of four hundred dollars a year. He had money to loan to his less fortunate neighbors and money to give to any good cause. Moreover, he had the advantage of living in the midst of good neighbors and close to store, church, post-office, and school. Think of it! This man out of debt, sure of his living year after year, with property and savings which entitled him to be regarded as moderately rich, and all from one irrigated acre; while neighbors with thousands of acres in wheat, and other neighbors with considerable areas in fruit, labored under loads of debt bearing high interest, or were sold out by the sheriff.

It is not to be contended that one acre of irrigated land is enough for the average family, nor that every man may expect to be as successful a gardener or as thrifty a manager as Mr. Cleek. When we see so remarkable a result as this we know that the personal equation must account for some of it. Nevertheless, the fact remains that small farms, under diversified and intensive cultivation by means of irrigation, would make the Sacramento Valley a paradise and enable it to support millions of people in a condition of rare independence and prosperity. Under such a policy this district alone would absorb enough surplus men to steady the Nation in some great crisis of the future. The development would demand not only farmers, but bankers, merchants, mechanics, and men and women of every profession.

The business opportunity presented by these conditions has at last attracted the attention of enterprise and capital, with the result that rich estates are being brought under irrigation, subdivided, and placed upon the market. Prices range from fifty to one hundred dollars an acre, so that small farms can be obtained by families of moderate means. On such farms, industrious settlers would be able to produce nearly all that they consume and have much to sell for cash. They may not realize at the beginning the most ideal social and economic conditions, for these wait upon the adoption of large public policies, but many a man will be able to improve his lot by joining the slender stream of settlement which is beginning to flow upon the irrigated lands of the Sacramento.

However, the great economic problems of this region will not be solved by private enterprise, alone. The work to be done is so vast that only the Government is capable of grappling with it successfully. Important steps have already been taken toward this end. Detailed surveys of the entire floor of the Sacramento Valley, together with careful studies of forestry conditions in its drainage basins, are under way. Ultimately, millions of dollars must be expended in the storage of flood waters, in the building of canals for irrigation and drainage, in the development of power, in the preservation and extension of the forests. The planning and directing of this work has fallen to the hands of Joseph Barlow Lippincott, one of the strongest characters who has been brought to the front by the national irrigation policy.

"I believe the Sacramento Valley offers the greatest undeveloped opportunity in Arid America, but I also believe that the problem involved in its reclamation is one of the most extensive and intricate that we have to deal with in Arid America," writes Mr. Lippincott. "I believe that it can be solved and that it will be solved, and I hope that the Reclamation Service will be able to lend material aid in its solution."

The golden age of colonization in the Sacramento Valley will come with the fruition of these plans.

The hope of that better and greater Sacramento Valley which is sure to unfold in the coming years was embodied in the life and teachings of a great man who passed from sight in June, 1905. This man was William Semple Green, familiarly and affectionately known to California for more than half a century as Will Green. He came with the Argonauts, but he had no eyes for the gold which absorbed their attention. From boyhood to old age, his one dream was to see the Great Valley watered, peopled, cultivated, and glorified by millions of independent homes. To make this dream come true he gave a lifetime of unselfish devotion. He lived to receive the assurance that all he had hoped and worked for would come to pass, though not to behold the reality with mortal eyes. He escorted the Congressional Irrigation Committees through the valley in June, 1905, and had the profound satisfaction of listening to the plans which

THE CONQUEST OF ARID AMERICA

the Government has made for a vast work of reclamation, and the even keener satisfaction of seeing these plans enthusiastically approved by that public which had for decades scorned the suggestion that irrigation alone could make the land of the Sacramento come into its own. His last words to the people of the valley, spoken at Red Bluff at the close of a banquet tendered to the visiting Congressmen, were these:

"For fifty years I have waited for this hour to come. I know now that I have not labored in vain. If I can but live to go up on Pisgah, and see this valley redeemed, and the home of God's chosen people, I shall be ready to lay me down and die."

A few days later, he passed on. But his work was done, and nobly done. His fame will live and grow with the valley for which he prayed as his life went out.

The San Joaquin Valley is even larger than its northern sister, which it resembles in nearly all fundamental respects. Indeed, the conditions of soil, climate, and productions are so nearly identical that they need not be rehearsed. There is one point of difference which is quite vital, however, and which has made itself felt in the history of San Joaquin. This is the fact that rainfall is appreciably less and that, as a consequence, irrigation is much more necessary. Hence, the small farm became popular many years ago and comparatively dense population has grown up in certain localities. The best example is the city and county of Fresno, in the heart of the valley.

EMPIRE STATE OF THE PACIFIC

Perhaps the earliest triumph of the new woman in this generation was that of Miss Austin and three other San Francisco "schoolma'ams," who founded the wonderful Fresno raisin industry. Investing their savings in a ranch and boldly venturing upon a culture in which few had faith, they demonstrated that raisins equal to those of Spain could be produced in California. They were rewarded with handsome profits, and thousands came to share in the benefits of their demonstration. All the evils that attend speculation in a single crop followed as a natural consequence, and brought a period of hard times. Unskilful irrigation without adequate drainage also wrought harm in various ways. But Fresno has largely outlived these misfortunes and the raisin industry gradually progresses to more stable conditions. On the whole, it played a wonderful part in the transformation of what was once but a poor stock range into a region of prosperous vineyards and beautiful homes.

The San Joaquin Valley has been much more fortunate than the Sacramento or any other part of California, in commanding private capital for the development of its agricultural resources. As a consequence, there are many opportunities for the settler to obtain good irrigated land on reasonable terms. J. B. Haggin and Lloyd Tevis invested millions of dollars in turning the waters of Kern River upon the rich delta of that stream. Bakersfield, which has become famous in recent years as the capital of an oil kingdom, is also the centre of the large district irrigated by the Kern. This is situated at the southern end of the valley. The rich Crocker estate has done a similar work on the Merced River, near the centre of the valley. In both cases opportunities have been opened to settlers which must otherwise have remained beyond their reach because of the large investment required to bring land and water together.

The San Joaquin has also been the scene of successful irrigation of another kind. Under what is known as the District Irrigation Law, large areas have been reclaimed by associations of landowners on the Tuolumne and Kings Rivers. The most important of these districts are the Turlock and the Modesto, on the former, and the Alta, on the latter. Industrious families of small means are making homes successfully in these localities, as well as in many other parts of the valley which are irrigated by smaller works.

The orange-growing district in the foothills of Tulare County, under the waters of the Tule River, are notably successful. Porterville is the centre of this district and seems destined to become an important city, as it is already a charming home-spot. Land and water are cheaper here than in the more famous orange districts of the south, but the profits of the industry are no less on that account. The higher improvement of the southern districts is doubtless due to the fact that a wealthier class of colonists was attracted there, rather than to superior natural advantages.

The valleys of the Sacramento and San Joaquin have been, and are yet, the grain-fields of the Pacific Coast. Many of their residents have bemoaned the fall in the price of wheat as the greatest of calamities. The truth is that for California it is the first of blessings. The fall in wheat prices has broken the land monopoly which kept labor servile and gave the most fruitful of countries to four-footed beasts rather than to men. Not until nearly all great ranches had been mortgaged to their full capacity, not until the failure of prices had made the debts intolerably burdensome and brought their owners face to face with disaster, was it possible to open the country for its best and highest uses. With the supremacy of wheat will go the shanty and the "hobo" laborer, to be followed in time by the Chinaman. In their places will come the home and the man who works for himself. Civilization will bloom where barbarism has blighted the land. There are localities where the cultivation of grain can be pursued, but the semi-tropical valleys of California were plainly intended for better things.

Irrigation, drainage, and cheap transportation are closely related as economic problems in the great interior valleys. William Hammond Hall, the former State engineer, has predicted that within fifty years the waters which rise in the mountains and meander through these valleys to the sea will all be utilized to moisten and fertilize the soil, and then be turned into canals, serving the double purpose of drainage and transportation. He claims that it is feasible, from an engineering stand-point, to construct such works, and to propel trains of freight-boats by electricity at a speed of six miles an hour. If this shall be done, the gain to the State will be beyond all calculation, provided the works be owned by the public. It is by no means an idle dream when considered in connection with ultimate California.

THE CONQUEST OF ARID AMERICA

The third field for future development is a vast region lying upon the eastern slope of the Sierra Nevada. This is so little known to the outside world that it may almost be named as Undiscovered California. It is reached only by lines of narrow-gauge railway running northwest and southwest, respectively, from Reno, Nevada. The northerly district is included in the three great counties of Plumas, Lassen, and Modoc. The country is distinctly arid, lying upon the western flank of the great basin formed by the Sierra Nevada and Wasatch ranges, which inclose portions of California, Idaho, and Utah, and all of Nevada. Here we find the real sage-brush desert— fertile, well-watered valleys surrounded by all the wealth of forest, mine, and natural pastures. The climate approximates much more nearly to that of New Mexico than to that commonly associated with the name of California. It is of the milder type of the temperate zone, favorable to the growth of such hardy fruits as apples, pears, peaches, and prunes. Up to this time, however, the chief products of the country are native and alfalfa hay, cattle, sheep, and horses. The sparse population is, perhaps, as prosperous as any farming community in the United States. This fact is mostly due to the vast extent of fine grazing lands surrounding irrigated valleys and to the herds of cattle and sheep which find their way to the farmers' hay-stacks from the ranges of northern California, southern Oregon, and western Nevada every autumn and winter.

The most important district in this region is Honey Lake Valley, lying eighty miles northwest of Reno. Here a new era has set in with water-storage for irrigation, small farms, and colonies planned upon the best ideals.

EMPIRE STATE OF THE PACIFIC

Cheap land, valuable surrounding resources, and a climate similar to that in which our race has flourished best, would seem to combine in favoring a large and rapid future growth.

The more southern body east of the Sierras lies chiefly in Inyo County. This is also at the early stage of development. The climate is milder, though still temperate rather than semi-tropical, than in the more northern counties. There are many beautiful valleys and an abundance of water, timber, and minerals.

Lack of railroad facilities and remoteness from large cities account for the backwardness of development in these attractive regions on the eastern slope of the mountains. They present to-day the finest field for development in California, and one of the finest in the United States. There can be no question that during this century they will become the homes of hundreds of thousands of people and the seat of a manifold industrial life.

The fourth field open to future conquest is a district which was formerly the most famous of waste places in America. In the earlier edition of this work, written in 1899, I spoke of this locality as follows:

"It is popularly regarded as an empire of hopeless sterility, the silence of which will never be broken by the voices of men. As the transcontinental traveller views it from his flying train it presents an aspect indeed forbidding. Neither animal life nor human habitation breaks its level monotony. It stretches from mountain-range to mountain-range, a brown waste of dry and barren soil. And yet it only awaits the touch of water and of labor to awaken into opulent life.* * * Much time will be required to overcome the wide and ingrained public prejudice against the Colorado Desert, but it will finally be reclaimed and sustain tens of thousands of prosperous people. It is more like Syria than any other part of the United States, and the daring imagination may readily conceive that here a new Damascus will arise more beautiful than that of old."

Six short years have passed, yet the dream has already come true. The very name of "The Colorado Desert" has vanished from the minds of men, and in its place we have a term which is synonymous with the highest productivity, with fat acres and fat cattle, with green fields, flowing waters, red-roofed farm houses, and rising towns. This name is "The Imperial Valley." In the summer of 1901, the waters of the Colorado River were turned upon this desert, and the most dramatic transformation ever seen in the United States quickly followed. Settlers and speculators rushed in to file upon the land and purchase water rights. Nearly a quarter of a million acres were thus acquired in the space of a few months, and the work of development went forward with furious energy.

While much of the land was taken under evil land laws which do not require actual residence or genuine improvement, nevertheless, over one hundred thousand acres were prepared for cultivation and planted to crops. Towns with schools, churches, banks, fine hotels, and all the conveniences of civilization, sprang up like magic. Railroads were quickly extended into the country, while telephone and telegraph supplied the means of quick

THE DESERT BEFORE AND AFTER.—Upper picture shows Colorado Desert (now Imperial Valley), California, as it appeared before irrigation—a brown waste of soil without vegetation. Lower picture shows the same land after irrigation, with two years' growth of cottonwoods.

communication. In a word, the picture which I drew from imagination in 1899 is reproduced on the face of the earth in 1905. And yet, what we now see is but the beginning of the achievement which will come in the future. This will be one of the most densely peopled agricultural districts in the world, and wealth which cannot now be estimated will come from the soil which, a few years ago, the average man would have scorned if offered a deed in fee simple for the whole domain. There could be no more wonderful example of the miracle of irrigation.

The work which has been accomplished by private enterprise in the rich delta of the Colorado River is one which was peculiarly adapted to national enterprise. Doubtless it would have been done by the Government if the private undertaking had been delayed by so much as two or three years. The National Irrigation Act became a law within twelve months of the time that water was first turned upon the desert. The region immediately attracted the attention of the national engineers, who proceeded to plan comprehensive works, of which we shall learn the details in subsequent pages. In the spring of 1904, a strong public demand arose for the inclusion of the Imperial region in the Government project. This demand was based to some extent upon the apprehension that the private system would prove unequal to the demands of the situation, but more upon complications which had arisen with Mexico, and upon the deep-seated popular antipathy to the private monopoly of water in an arid land. The movement encountered temporary failure because the Government and the private owners could not agree on the purchase price of the irrigation system which had produced such sudden and dazzling results in the development of the region. The price recommended by the settlers and accepted by the Company, was three million dollars. Whether this figure was reasonable or exorbitant can only be determined by future events, but there can be no possible doubt that the final outcome will be a single comprehensive work of reclamation, from the Grand Canyon of the Colorado to the Gulf of California, controlled by the United States of America. Nothing short of this is worthy of the opportunity.

The town life of California, considered from the standpoint of the man who contemplates going West, is not strikingly different from that in any other part of the United States. The growth of towns is out of all proportion to the development of the surrounding country. This is due rather to an influx of Eastern people than to any local tendency to desert the rural districts. If in certain restricted localities young men are leaving the farm for the city, the loss is more than made good by the number of newcomers who are seeking homes in the country. The fact remains that urban growth is strongest because the majority of those coming from older sections of the United States seek to make places for themselves in the larger centres of population.

The commercial and political centre of California is, of course, San Francisco. The State's front door is the Golden Gate. The metropolitan community, which may be defined as Greater San Francisco, includes all the

TOWN BUILDING IN THE DESERT.—Upper picture shows beginning of Imperial, California, March, 1901. Lower picture, one side of same street four years later.

EMPIRE STATE OF THE PACIFIC

cities and towns about the bay. And though this community is little more than fifty years old, it is already about the size of Boston. There is apparently no reason why it should not, in the course of the long future, become as large and important in every way as New York itself. Never was there a better foundation for a mighty city than in San Francisco and its environs—the mountain-sheltered harbor, with wide expanse of deep waters; the sloping shores running back to foothills and mountains; the noble bay stretching inland, with rivers navigable far into the interior; half a continent behind it, and before it the measureless possibilities of foreign worlds. When to these substantial considerations are added the charms of climate and scenery and the social and educational advantages which have arisen, and must arise much more in the future, from the wise use of private and public wealth, one might be excused for dreaming of San Francisco as the focal point of civilization in coming generations.

Of the cities about the bay, Oakland is second in importance to San Francisco, to which it sustains about the same relation as Brooklyn to New York. Like Brooklyn, it is becoming important in manufacture and commerce, having a magnificent water frontage where goods may be moved to east-bound trains directly from the ships, instead of having to be ferried across the bay, as in the case of the sister city. Alameda is a delightful residence suburb. Berkeley is the seat of that institution which is the pride of the people, the University of California. Among the numerous and beautiful towns near San Francisco, Palo Alto enjoys a special prominence as the site of Leland Stanford Junior University. The growth of all the communities clustering about San Francisco, while not sensational in recent years, is constant and substantial.

Sacramento is the chief city of the valley of that name. It contains about thirty thousand people and is the capital of the State. While its relative importance is by no means the same as in the early mining times, it is a beautiful and influential community, and destined to be more so as its surrounding resources are developed.

Of the cities of the San Joaquin, Stockton is most advantageously located. It stands at the head of navigation on the San Joaquin River, in the midst of a rich agricultural region. It should be, as it is now becoming, the seat of many industries. It is difficult to place limits upon the reasonable growth of Stockton whenever the policy of irrigation, of the subdivision of lands, and of diversified farming, shall be made to take the place of dependence upon rainfall, of large holdings, and of the single crop. It already enjoys cheap freight and passenger transportation to San Francisco, by lines of steamers which cover the distance in a night. The possibilities of this traffic are immense and the advantages which it confers upon Stockton as a commercial and manufacturing city in connection with agricultural development are well-nigh incalculable.

The chief cities of the south are Los Angeles and San Diego. The growth of the former is amazing.* Its

* Los Angeles had a population, by the U. S. Census.

In 1850,	1,610	In 1880,	11,183
1860,	4,385	1890,	50,395
1870,	5,728	1900,	102,479

present total is not far from two hundred thousand, and no one familiar with its history would be much surprised to see it approximate half a million by the date of the next national census. The secret of its vitality is not found on the surface. It is not due to its commerce, for it is twenty miles from the sea and without a natural harbor, though even this deficiency is now being supplied by public enterprise. Neither does its trade with the interior account for its constant and ever-increasing prosperity. The superlative charm of Los Angeles and the region surrounding is social rather than economic. Men and women desire to live where they may realize their highest possibilities as social beings. And Los Angeles has become the metropolis of a district which satisfies this instinct more fully than any other part of the United States. People come originally as tourists to spend a few blissful weeks in the winter. Many of them return as colonists and throw themselves into the life of the place with all the zeal of converts, thereby swelling a tide of public spirit which is already indomitable. Their confidence in the future is boundless, and they "pull their weight,"—all of them. That it is which has made Los Angeles.

In the most remarkable book ever written about California, " The Right Hand of the Continent," Mr. Charles F. Lummis makes the following striking statements concerning the growth of Los Angeles:

" Not one city in the United States which was no larger than Los Angeles in 1890 is larger now; not one city which was no larger in 1880 is larger now. In other words, not a single city in the Union has overtaken Los Angeles in rank by population.

THE CONQUEST OF ARID AMERICA

But in these two decades, Los Angeles has outstripped 99 American cities which were numerically larger in 1880 ; and in one decade has passed 19 cities that were numerically larger in 1890. In 1880, Los Angeles was the 135th city in the Union in population. In 1890 it was the 56th. In 1900 it was the 36th. There are now 35 cities in the United States larger than Los Angeles ; but only 13 cities have *gained as many people* in the ten years from 1890 to 1900."

And he shows that the cities which scored a larger increase in that time were as follows: New York, Chicago, Philadelphia, St. Louis, Boston, Baltimore, Cleveland, Buffalo, Pittsburg, Detroit, Milwaukee, Newark, and Indianapolis.

Southern California is becoming the playground of the Republic, and Los Angeles is its capital. But those who come to play remain to work—to build hotels and office buildings, to establish railroads and factories, to develop the rich natural resources of the country. No ordinary rules explain its past growth or set limits to its future expansion. It has been, and it will be, a law unto itself.

The case of San Diego is somewhat different. A great city might grow up upon its site for precisely the same reasons that great cities grew up at Boston and New York. San Diego has the only natural harbor in California south of San Francisco. It is the nearest American port to the Western outlet of the Panama Canal. It occupies a strategic position with relation to the commerce of Central and South America, Australia, and the Orient. It is the logical terminus of a transcontinental railway which would make the shortest line across the continent to some port on the South Atlantic Coast of the United States. Such a railway would have the lowest grades and the most certain immunity from snow blockades. It ought to be the seaport and trading centre for Arizona—which is another South Africa—and the whole southwest. If nature ever planned the site of a great city, it did so where the encircling shores rise from plain to hill, from hill to mountain-range, about the lovely bay of San Diego. The city has a population of over twenty-five thousand and is growing rapidly.

It is conceded that San Diego enjoys the best climate in the United States. Those who visit it in winter declare that that season is more delightful than its perfect summer; and those who visit it in summer declare that it is then even better than in winter. And both are right! The best season in San Diego is the one you last spent there, whether it happens to be winter or summer. The city lies between those vast mysteries, the desert and the sea, and its atmosphere is a charming blend of both. These extraordinary climatic conditions must in time give the place and its surroundings an enviable pre-eminence as a popular resort all the year round. Eastern people will go there to escape the cold in winter, while thousands will flee from the hot interior to enjoy its cooling breezes in summer.

The future of California will be very different from its past. It has been the land of large things—of large estates, of large enterprises, of large fortunes. Under another form of government it would have developed a feudal system, with a landed aristocracy resting on a basis of servile labor. These were its plain tendencies years ago, when somebody coined the epigram, " California is the rich man's paradise and the poor man's hell." But later developments have shown that whatever of paradise the Golden State can offer to the rich, it will share, upon terms of marvellous equality, with the middle class of American life. Over and above all other countries, it is destined to be the land of the common people. This is true, because, owing to its peculiar climatic conditions, it requires less land to sustain a family in generous comfort. For the same reason, cheaper clothing and shelter, as well as less fuel, suffice, while it is possible to realize more perfectly the ideal of producing what is consumed. Moreover, it is a natural field for the application of associative industry and the growth of the highest social conditions. Indeed, the country has distinctly failed as a land of big things, and achieved its best successes in the opposite direction. Its true and final greatness will consist of the aggregate of small things—of small estates, of small enterprises, of small fortunes. Progress towards this end is already well begun. It must go on until the last great estate is dismembered and the last alien serf is returned to the Orient. Upon the ruins of the old system a better civilization will arise. It will be the glory of the common people, to whose labor and genius it will owe its existence. Its outreaching and beneficent influence will be felt throughout the world.

CHAPTER II

THE NEW DAY IN COLORADO

THE old day in Colorado was the era of frontier barbarism. The glitter of Pike's Peak gold drew throngs of adventurous folk who toiled across the plains of Kansas and Nebraska in wagon-trains that they might speculate in the mysterious possibilities of a new country. They were not home-builders, but fortune-hunters. Wherever they found placer gold rude settlements sprang up.

In the mean time the cattle industry began to contend with Indians and buffalo for the possession of the grazing lands which sloped away from the Rockies, and the necessity of a base of supplies planted the seeds of a few permanent towns, such as Denver and Pueblo. These were mere clusters of rude homes and stores which seemed to hold out scant promise of future importance. The Indians were numerous and troublesome, and the life of the pioneers was spiced with danger. Though the country belonged nominally to Kansas, there was but the slightest pretence of civil government. Practically the only authority was that exercised by organizations of citizens, who brought horse-thieves and murderers to speedy justice upon the most convenient tree.

In 1861 Colorado became a Territory, and was then

THE CONQUEST OF ARID AMERICA

able to deal more effectively with the Indian, who was the common enemy and an obstacle to settlement and development. There was little in these early conditions to encourage the hope that a great and populous State could be established amid the mountains and plateaus. Mines, cattle, and border traffic were not alone sufficient for the making of civilization. Beyond these crude industries the future was speculative. The country was unexplored, the resources undeveloped, the conditions untried. The transformation which swiftly followed upon this period of doubt converted the frontier community into one of the most brilliant and promising of American States.

The dawn of the new day was heralded by the whistle of the locomotive. The dissolution of the Union armies had turned the faces of many thousand veterans towards the trans-Missouri region, and of these Colorado received its full share. The wonderful era of railroad-building—perhaps the most dramatic page in all our industrial history—had just begun. These circumstances conspired to give a new and powerful impulse to the territory at the base of the Rocky Mountains. Large capital joined hands with the increasing stream of immigrants, and Colorado entered with amazing vigor upon a stage of real and far-reaching development. More important than the finding of gold was the discovery of the fact that the highest forms of agriculture would flourish with the aid of irrigation. When this had been demonstrated by the pioneers there was no longer doubt about the future greatness of the State or the character of its civilization. Denver and a few other settlements began to take on the appearance of permanency, and

THE NEW DAY IN COLORADO

even to exhibit the signs of coming refinement and power.

The settlers of Greeley inaugurated large irrigation enterprises and planted seeds from which the finest civic institutions were to grow. General William J. Palmer and his friends, anticipating the commercial value of climate and scenery even before the industrial economy of the community was established, laid out Colorado Springs, at the foot of Pike's Peak, and began to make Manitou and the Garden of the Gods ready for future thousands of health-seekers and tourists. Pueblo quickly felt the importance of its position on the banks of the Arkansas at the gateway of the mountains, and developed rapidly in population and business. The daring conception of a railroad to parallel the Rockies and open communication with Mexico, or to scale the giant peaks and penetrate the wilderness which lay beyond, took possession of General Palmer's mind and furnished the hope of further extraordinary developments.

Thus the decade between 1870 and 1880 saw the rise of Colorado to a place of immense promise and of important achievement, and in 1876 the nation signalized the centennial of the Declaration of Independence by bestowing the rich privilege of sovereignty upon the newborn commonwealth.

The Colorado of to-day contains a population of more than half a million. It is marvellously fortunate in its railroad development, having twenty-four separate lines, which maintain over five thousand miles of track, penetrating nearly every part of the State. Its mines of precious and base metals—very largely the former—yield an annual income of nearly fifty millions.

THE CONQUEST OF ARID AMERICA

Its two million acres of irrigated land add forty millions more to the annual industrial product. Manufactures, including smelting and refining works, produce goods to the value of one hundred millions. Other business transactions, represented by the commercial and professional classes, represent considerably more than one hundred millions each year. The live-stock industry is difficult to estimate, but adds very largely to the yearly production of wealth.

Such are the results wrought out by the labor of a single generation upon the raw resources of a new State. Before glancing at the people who have organized such an economic life in so brief a space of years, and at the institutions they have created, it is important to consider the material foundation on which they have built.

Colorado owes something to its scenery, much to its climate, yet more to its mines. The first of these made it widely known as one of nature's wonderlands. The second was a prime factor in attracting population. The third poured a large and continuous stream of wealth into the hands of the people, and a little further on we shall see how loyally this has been used for the benefit of the State. The grandeur of the scenery and the charm of the climate are both matters of popular knowledge. Neither is peculiar to Colorado, for both are characteristic of the arid region as a whole. But nowhere else do the ordinary paths of travel lead through so grand a scenic region as in Colorado, nor has any other locality been as fortunate in the energy and intelligence bestowed upon the work of making this phase of its attractions widely and favorably known.

The Colorado climate is the product of high altitude

THE NEW DAY IN COLORADO

and aridity. Denver is one mile above the level of New York harbor, and much of the inhabited portion of the State is even higher. The result is a rarefied atmosphere very exhilarating in its effects and extremely favorable to persons suffering with certain kinds of diseases. Summer and winter are almost equally delightful, though presenting great extremes of heat and cold.

Of the mineral wealth it is needless to say more than that it increases its annual output with regularity, and that there is every reason to suppose that much the greater part of it yet remains to be discovered and developed. It will be a permanent resource of the highest utility, since most of it is directly converted into money at the local mints. While the energies of the mining industry are chiefly centred upon the search for precious metals, the country is endowed with the greatest variety of mineral riches. These include nearly all the base metals, such as copper, lead, and iron, as well as coal, oil, precious and semi-precious stones, granite, marble, onyx, and sandstone. These materials exist in the greatest profusion, but must lie mostly unused until the population largely increases.

In considering the matter of agricultural development, it must be remembered that Colorado is the crown of the continent. Its lofty mountain-peaks cut the rainfall and melting snows in twain, sending one part to the Pacific and the other to the Atlantic Ocean. The same influence makes a radical division in climate, productions, and the character of agriculture. Irrigation development naturally began earliest where streams could most easily be diverted. This was on the high plateau which slopes eastward from the foot-hills and merges into the Great Plains of Nebraska and Kansas.

For a period of nearly twenty years, beginning in 1870, canal construction and the settlement of lands were actively carried on in this part of the State. The scene of action was principally in the valleys of the Cache la Poudre, the Platte, and the Arkansas. Here the farms are of large size for an irrigated region, though the present tendency favors a smaller unit. These districts have taken on a new growth of late, with the prosperity of the State and the country. The products are diversified and largely disposed of in the home market. In the upper Arkansas Valley, where the foot-hills furnish shelter from the high winds prevailing at certain seasons, fruit-culture has been notably successful. Prices of unimproved lands on the eastern slope range from twenty to fifty dollars per acre, while cultivated lands are valued at one hundred dollars an acre and upwards, according to the extent of improvements and location with reference to cities or large towns. The glimpse we have had in an earlier chapter of the agricultural industry of Greeley Colony may be accepted as true of the entire region east of the mountains, for Greeley has been the model to which other districts have looked for inspiration. The experimental farms which surround the agricultural college at Fort Collins undoubtedly represent the highest type of irrigation results in this part of the State. In the Arkansas Valley the altitude is lower and the climate more favorable for small farming and fruit-culture.

The San Luis Valley is an elevated plateau lying between parallel mountain-ranges in the southern and

IRRIGATED APRICOT ORCHARD, NEAR MONTROSE, COLORADO.

THE NEW DAY IN COLORADO

central part of the State. Here a vast expenditure has been made for irrigation works, but the earlier efforts at settlement were disappointing. The explanation is not to be found in the altitude, which is from seven thousand to eight thousand feet above sea level, for the country makes a wonderful yield of grain and of vegetables, as well as of apples and small fruits. While it is true that many of the earlier settlers made failures through different causes, yet there are instances of the greatest prosperity on the part of others. Numerous experiments have demonstrated the adaptability of the soil to the raising of sugar beets, and it is probable that this valley will share with other portions of the State in the prosperity attendant upon the beet sugar industry. A striking example of prosperity is seen in the thriving communities of Mormons. The industrial system which we have already studied in connection with Utah produces the same good results in the San Luis Valley. In view of these facts it must be assumed that the locality will eventually be thickly settled and sustain thousands of prosperous people. Land and water may be obtained more cheaply here than anywhere else in Colorado and there is a good market for the products of the soil. The costly preliminary work of reclamation has been well done in advance. A labor colony, founded upon wise plans, backed by sufficient capital, and inspired and managed by skilful leadership, would solve the problem of colonization for the San Luis Valley, while furnishing work and homes for those who need them. The Mormon communities are practically of this character in the beginning.

The western slope of Colorado constitutes a region entirely distinct. From a casual glance at the map it would be inferred that about two-thirds of the State consist exclusively of mountains, and are therefore unfitted for settlement. The truth is that there are many beautiful valleys of varying size and elevation, and that these are destined to sustain the most interesting and profitable agricultural districts of Colorado. Unlike the eastern slope, there is here more water than irrigable land— a condition almost unique in the arid region. The valleys are so protected by the mountains which inclose them upon either hand as to have a climate of their own. This is perceptibly influenced by the warm winds which make their way from the Gulf of California through the canyons of the Colorado river. These conditions are extremely favorable for the culture of the most delicate fruit and for the diversification of general crops. The principal rivers of the western slope are the Grand, the Green, and the San Juan. These are fed by the prolific snows of the higher Rockies, and carry a strong and turbulent flow of water throughout the year. They are not always readily diverted, however, as their channels have been deeply cut through the rocks and soil, and the stream often flows below the level of the tract to be irrigated. This makes it necessary to elevate the water in many instances by pumping machinery, which can be operated cheaply by the power of the stream itself, or by the use of coal, which in many cases is found close at hand.

The best example of the possibilities of the western slope is seen in the neighborhood of Grand Junction, where two splendid streams—the Grand and the Gunnison—join forces and flow westward to their meeting with the Green river across the Utah boundary. Here the valley opens out into a broad desert, with foot-hills, or *mesas,* marking the rise to the mountain masses which line the horizon on either hand. To the eye of the traveller who has just come through the awe-inspiring scenery of the mountains and narrow upper valleys, nothing could be less promising than the brown waste of arid soil which he beholds upon approaching Grand Junction. The scene is one of utter desolation, for even sage-brush and mesquite are absent from large portions of the landscape. The roaring river hurrying down the slope seems to mock, with hoarse laughter, the unfruitful soil, which stretches away from its banks in silence and in sunshine. But if the traveller leaves the train and rides out a few miles upon the desert he will quickly interpret the mystery of these conditions. Wherever the water has been married to the soil, prolific fields and orchards have sprung from the union—such fields and orchards as may be rivalled as yet only in semi-tropic California. The favorite size of farms is from ten to twenty acres, or only about one-fourth or one-eighth of the average area of farms on the eastern slope of Colorado.

Fruit-culture chiefly claims the thought and energy of the people in this locality, and it is very profitable. Peaches are the leading product, and they are wonderful for flavor, size, and beauty. A local festival is "Peach Day," when people come from all directions to feast upon the free bounty of Grand Junction. Lands are held high, ranging from fifty to one hundred and fifty dollars per acre, though they were but recently public property and of no value until irrigation facilities had been provided. The excuse for these high prices is the fact that orchards in bearing frequently earn one hundred and fifty

THE CONQUEST OF ARID AMERICA

dollars and upwards per acre each year. This is due in part to the marvellous quality of the fruit, and in part to the extensive home markets offered by mining camps in the mountains, and by large towns such as Denver, Pueblo, and Colorado Springs. In view of the severe limitations which nature has placed upon the territory suited to the highest culture of delicate fruits, and of the steady growth of the consumers in mountain districts and large towns, there is, perhaps, good reason to hope that profits will be well sustained for a long time to come.

These conditions make the western slope choice ground for settlement. They are by no means limited to the lower valley of the Grand, but exist in the numerous smaller districts scattered through the mountains in the western and southwestern part of the State. On the social side the possibilities of the country have not been much developed, as there has been a lack of organized effort in settlement. But the extraordinary fertility of the soil, the extent of the water supply, the proximity of mining camps, and the charm of the climate must sometime combine to lend a powerful impulse to the highest development of these favored valleys.

The scenery presents not merely pictures, but pictures that are painted and tinted and wrought into fantastic shapes. To the ever-changing aspect which the mountains, buttes, and *mesas* gain from light and shadow, from sun and cloud, new and strange beauties are added by the reds, pinks, yellows, and grays of soil and rock. From the vivid cliffs and bluffs which stand guard upon river banks to the purple and shadowy peaks which lift their pointed heads on the utmost horizon, the scene is one of

VIEW LOOKING UP GUNNISON RIVER NEAR UNCOMPAHGRE PROJECT, COLORADO.

THE NEW DAY IN COLORADO

such beauty and grandeur as may be felt, though not described.

Such are the materials of Colorado. Let us look now at the people and their civilization.

Intense local patriotism is a well-recognized western trait, but in Colorado it amounts to a religion. We have seen how the progress of California was impeded by certain elements of its population having no sympathy with its higher ideals, no pride in its best achievements. If there is such an element in Colorado it is unseen and unfelt in the larger life of the State. The community is dominated by a spirit of aggressive enterprise which recognizes no impossibilities, harbors no doubts of the future. This is the explanation of what we may fairly call—in view of the brief time consumed in its evolution from conditions essentially barbaric—the splendor of Colorado civilization. It is this which created Denver, almost the fairest of American cities; which made Colorado Springs the centre of wealth and refinement; which blackened the sky of Pueblo with the smoke of a young Pittsburg; which planted Leadville among the clouds; which placed a steam ladder against the dizzy summit of Pike's Peak; which carried the iron highway of commerce through gorges and mountain-passes; which turned rivers out of their courses that barren soil might blossom with the homes of men. This high public spirit is seen in schools, colleges, clubs, public buildings, and improvements—above all, in the homes.

It has been the policy of those who have taken riches from the mines to invest them in developing the State's resources and in beautifying its cities and towns. In this respect the spirit of Coloradans presents a sharp contrast to that of many who grew rich in California, and of most of those who received the enormous wealth coined from the resources of Nevada. In the latter instance the beneficiaries of the mines did not even make their homes in the land which raised them from poverty to affluence. But the men of Colorado have been proud of their devotion to the commonwealth which they created, and have striven by every means in their power to keep it moving along the upward path. In the erection of fine public and business buildings and of palatial homes, in the extension of railroads and irrigation canals, in the increase of banking capital, and, above all, in the pursuit of daring mining operations, their enterprise has been unequalled by that of any other western community. Foremost among those who inaugurated this policy at the risk of their fortunes was the late H. A. W. Tabor, whom Denver and Colorado should always hold in grateful remembrance.

But there is another side to the picture. The tendencies of Colorado civilization are not wholly in line with the best ideals of the arid region. Viewed from this stand-point, its institutions are in a measure disappointing. The marvel of Denver's growth and the beauty of its homes and business districts should not blind us to the fact that it is essentially like the great cities of the East. It is, in a word, another case of "progress and poverty." The equality which marked its early life has diminished in proportion to the growth of the population and the increase of wealth. The rise of land values has made it more difficult for the many to own their homes, and has increased the wealth of the land-

THE NEW DAY IN COLORADO

lord class. All the evils which grow from the conditions of life in a large city are rife in Denver.

These are not the natural economic tendencies of a country founded upon irrigation. They are not such as we have observed in localities where irrigation has been so nearly the dominant influence as to shape institutions. The explanation is found in the influence of mining speculations which, diffused like the atmosphere, breed a cheerful but demoralizing contagion: also in the early tendency to adopt a comparatively large farm unit. These two forces have operated to produce very different results from those flowing from the Mormon land policy, which we saw in the Salt Lake Valley; or from those which grew in consequence of irrigation in the San Bernardino Valley of California. Large portions of Colorado are admirably adapted to the development of the best social conditions—of those conditions which make for a permanent and growing body of landed proprietors; for the multiplication of little towns rather than a concentration of people in congested centres; for the application of the associative principle in connection with industrial and commercial affairs. It is gratifying to be able to record that the latter currents of thought in Colorado seem to show the effects which might be expected to result from its environment.

More and more the State asserts its authority in the control of irrigation works and practice. The farm unit grows smaller, and intensive cultivation finds more followers. By enormous majorities the people pronounce in favor of party platforms which demand the public ownership of public utilities. Equal suffrage and the presence of women in the legislature mark the progressive temper of the body politic. On the whole, there is much reason to hope that the social achievement of the next generation in Colorado will be equal to the material achievement of the last.

CHAPTER III

THE PLEASANT LAND OF UTAH

THE industrial system of the people who compose three-fourths of the population of Utah has been considered in connection with typical institutions of the arid region in earlier pages. It remains to speak of the physical aspects of the newest of American States.

Standing on the summit of Capitol Hill in Salt Lake City, one may take in the entire range of Utah's resources, developed and undeveloped, in a single sweeping glance.

At one's feet lies the mountain metropolis, with the stately temple of native granite supporting the golden figure of the Angel Moroni on its culminating turret, and beside it the odd-roofed tabernacle, like an enormous turtle basking in the sun. Below, the miles of city streets stretch southward—a huddle of business blocks in the centre; a series of garden-homes hidden by leaves and blossoms on either hand. Still farther out the generous city lots expand into little farms of ten or twenty acres, exemplifying the prosperous irrigation industry, which is the corner-stone of the commonwealth. Far down the valley the smelters send up their black smoke to the sky—emblem of the mining industry. At the lower end and on the sides of the valley lies an ex-

THE CONQUEST OF ARID AMERICA

panse of arid land in its natural desert state, typifying alike the conditions encountered by the pioneers and the present aspect of a vast proportion of Utah. On the left, one sees hastening down the canyon the roaring creek which watered the first crop ever planted in these valleys; on the right, the glistening expanse of the famous inland sea. And inclosing all, the mountains—treasure-house of precious metals, of coal, of iron, of timber, and of the snows and waters which fertilized the desert and made it blossom with civilization.

Here in a single picture is all of Utah—town and country, farm, workshop, mine, shrines of religion, and play-grounds of wealth and leisure. If the human eye might look beyond the brown barriers, which now intercept the view, to the very boundaries of the State, it would see nothing more than it sees from Capitol Hill, for Utah is a succession of mountains, of desert valleys, and of crystal streams, and scattered over it all is the wealth of the mine and the sleeping potentiality—here and there partially awakened—of the home, the field, the orchard, and the workshop. It is a pleasant and a sunny land, unforgotten by the most casual traveller who has crossed it and well loved by those who claim it as their home. It is easy to understand the feelings of the little Utah boy who tired of the World's Fair in a very few days and begged, with tears in his eyes, to be taken back. Asked if there were not plenty of interesting sights in Chicago, he replied, "Yes, but I can't see no mountains!"

Utah has a population of about a quarter of a million. Though this is but one-half as many as Colorado, and one-fifth as many as California, the new State approaches more nearly to the ideal of a self-supporting community than either of its neighbors. The bulk of its population has been trained in the policy of industrial independence from the time of its earliest settlement. We have seen how this was accomplished with little capital except that which was taken from the soil. The fortunate results may now be observed in an industrial life which is remarkably diversified for a community so new and remote.

Very much the larger portion of the population may be seen in a railroad ride of two hours, from Provo through Salt Lake City to Ogden. This ride takes the traveller through Utah, Salt Lake, and Weber valleys, which were the first to be reclaimed, and must always contain the densest population. The original advantage of this now splendid district was its abundant water supply, flowing in numerous streams from high mountains near at hand. To this advantage later development added the presence of important railroad systems and the proximity of rich mines of precious metals. The growth of other portions of the State, which must be large and constant, can only confirm the supremacy of the communities which have grown up near the shores of the Great Salt Lake. These are alike the commercial, political, and religious centres of Utah, to which all the sources of material wealth must be tributary.

The natural resources of Utah, as in the case of all the States of the mountain region, are wonderfully diverse, though in the infancy of development. The annual output of gold, silver, copper, and lead is now about ten million dollars, and is constantly increasing. The mining industry is thus a large contributor to local wealth, supplying employment to thousands of laborers, furnish-

THE CONQUEST OF ARID AMERICA

ing a home market for the products of the farms, and giving constant encouragement to the extension of the railroad system. The work of discovery and development in new districts steadily progresses, and the economic value of mineral resources must grow with every passing year. Utah is somewhat deficient in forests suitable for timber, but is abundantly endowed with coal, iron, and water-power, which are the raw materials of manufacture. The development of water-power in connection with electricity has begun in earnest and will be a factor of high importance in the future. This is accomplished by damming streams which flow through mountain canyons in the immediate neighborhood of large towns. This requires the transmission of electricity for a distance of only a few miles, owing to the fortunate natural conditions. The State is also rich in fine building stone, which includes beautiful marble and onyx.

The climate of Utah is that of the milder temperate zone, and during large portions of the year is thoroughly delightful. Ploughing begins earlier than in eastern localities of similar latitude. The spring days are showery and windy, but the first warm breath of approaching summer is usually felt by the last of April. From May until November there is little rain. The thermometer climbs high during the summer days, but the heat is not oppressive, owing to the dryness of the air. Mountain breezes, sweeping down through the numerous canyons, make the nights delightfully cool. In Utah it is the custom to run irrigation waters through the streets of cities and towns during the summer, and the music of these numerous babbling streams is a pleasant feature of the country, and apparently of considerable effect in mitigating the heat. The long autumn, extending frequently into December, is the most charming season of the year. The winter is usually brief, but accompanied by considerable snow even in the valleys and a very heavy precipitation in the surrounding mountains. On still nights the thermometer sometimes goes well below zero. The extreme southern portion of the State, locally known as "Dixie," is much milder, indeed verging upon the semi-tropical, and permitting the culture of figs, almonds, and English walnuts.

The agricultural industry of Utah presents some odd contradictions. It is more diversified, and therefore more completely self-sustaining, than that of any other western State. Farms are smaller and less incumbered with mortgages, and the people may be said to live generally in easier circumstances than the occupants of the soil in any other part of the United States.

On the other hand, it is not here that we find the best methods of irrigation and cultivation, nor of packing and marketing the crops. The high intelligence and persistent effort which placed certain communities in Colorado and California at the head of the list in their respective lines of production are wanting in Utah. The fruit possibilities of the country have been especially neglected until recently, so that newly settled portions of Idaho have easily surpassed Utah localities which had the advantage of more than a generation in time. Of late years there has been a marked improvement, resulting from a State Board of Horticulture, from the influence of the Agricultural College at Logan, and from the infusion of a considerable element of new settlers.

THE CONQUEST OF ARID AMERICA

Over half a million acres of irrigated land are in actual cultivation, while nearly twice that number are under canals now completed or in process of construction. Nearly one hundred thousand acres are cultivated in grain crops without irrigation. These are mostly situated north of the Great Salt Lake, where the rainfall is heaviest. The total amount of cultivated land is, however, only about one per cent. of the area of the State. According to the best local authorities, something like six times as much land as is now irrigated can be brought under cultivation by these methods when the water supply is utilized. Here is a large field for the growth of population.

The territory available for settlement is well distributed throughout the State. The country immediately surrounding the three large towns of Ogden, Salt Lake, and Provo is compactly settled, yet better methods of utilizing the water supply will enlarge the area of cultivation even in those districts. The beautiful country lying immediately north of Great Salt Lake, and watered by one of the largest irrigation systems in the West, is still largely open to settlement. Here the fruit industry is rapidly developing in connection with general farming and stock-raising. In this locality unimproved lands sell for prices ranging from thirty to fifty dollars per acre, while the annual water-rental is two dollars and a half per acre. The construction of new irrigation systems in the large deserts south of the lake, in central Utah, has been actively carried on during the past five years. Here much government land is open to entry, but the settler must purchase water-rights from canal companies. This item of cost should be added to

A TYPICAL MOUNTAIN STREAM IN THE ARID REGION—MOUTH OF ECHO CANYON, UTAH, SHOWING WEBER RIVER

THE PLEASANT LAND OF UTAH

the price of the land. In this locality unimproved lands with water cost from ten to thirty dollars per acre. The raising of grain and hay is profitable because of the demand which the stock industry furnishes for these products, while the culture of peaches, apricots, apples, and prunes seems promising. These fruits have been raised successfully for forty years in the more sheltered valleys and foot-hills of central Utah, and the later orchards are being gradually extended farther out upon the desert.

For years settlers, miners, and speculators have looked with eager eyes to the Uinta country, surrounded by the mountains of that name and lying directly east of Salt Lake City. Here a great Indian reservation has been thrown open to settlement. The opening of this new district was the occasion of a rush of land-hungry people, in which the Mormons took a conspicuous part. They will undoubtedly apply their well-known successful methods of colonization in such localities as they are able to control. Settlers will be organized into companies constructing their own canals by combined labor and dividing the farms and village lots under an equitable arrangement. The Uinta country is rich, not only in agricultural land, but in minerals, timber, building stone, asphalt, and other useful resources. Its value has long been appreciated, but railroad facilities were lacking. These will now be supplied by a new line building west from Denver. Several towns are certain to spring up and attain importance at an early day. The deserts of eastern Utah within reach of the Green River, and in southern Utah in the neighborhoods of the Colorado and Virgin Rivers, have but begun to feel the influence of

THE CONQUEST OF ARID AMERICA

modern enterprise. The costly works necessary to their reclamation will doubtless come as the pressure of settlement increases.

Utah's pre-eminence in the land of irrigation is due to historical considerations rather than to the excellence of its canal systems or to the superiority of its laws and customs. In the latter respect it is distinctly disappointing. The pioneers turned the water from the most convenient streams by the crudest devices, and with no thought for any grand and enduring scheme of engineering. Canals were often duplicated many times over in a single valley, wasting precious water, injuring the soil, and unnecessarily restricting the area of settlement. The evils of the irrigation system hastily constructed by the pioneers are now seen and felt; yet the early appropriators of the mountain streams are so tenacious of what they consider their rights as to render the reform of the laws, the reconstruction of canals, and the readjustment of irrigation customs to meet the conditions imposed by the pressure of population, extremely difficult. Efforts to establish a plan of State supervision which would provide for the measuring of water and its just apportionment among irrigators—a system which is the first and last essential of peace and progress in an arid land—have been repeatedly frustrated by the unreasoning jealousy of the older settlers. In recent years, however, this opposition has yielded to higher conceptions and Utah now has a good administrative system which is gradually reconstructing the basis of its water rights.

For fully forty years Utah irrigation was held in the hands of small local companies composed exclusively of the land-owners. Works were built by the common labor

THE PLEASANT LAND OF UTAH

of the community, and the repairs and improvements made in the same way. The first important departure from this policy came with the construction of the bold and expensive canals of the Bear River Irrigation Company, which have reclaimed a large area lying between the Great Salt Lake and the Idaho boundary. These works also supply domestic water to the city of Ogden and furnish power for electrical purposes. The Bear river canal is one of the most notable works of engineering in the United States, ranking at least second, if not first, among irrigation systems in this respect. Not far from two million dollars of eastern and foreign capital is invested in the enterprise. The work exhibits almost every phase of irrigation — engineering, including canals cut into solid canyon walls, tunnelled through mountain sides, as well as iron flumes and notable diverting dams. Other private water systems followed the Bear river development. The most important of these are the storage enterprises at Mount Nebo and in the neighborhood of Sevier lake. Both of these utilize the flood waters of the Sevier river, which is one of the largest streams in the State.

No other community in the West will deal with more interesting irrigation problems in the future than Utah. The conflicts between the policies of public and private ownership cannot be avoided, since both are represented in systems which lie side by side. In districts where settlement is furthest advanced and canal systems the oldest, the crying necessity for the reconstruction of works and the application of a rigid public supervision must soon be answered. Coincident with the settlement of these questions will be the gradual evolution of better agricultural and horticultural methods.

The construction of the San Pedro, Los Angeles & Salt Lake Railroad, familiarly known as "the Clark Road," is the most important event in Utah history since the building of the Union Pacific system. It will open a large mineral, agricultural, and stock region south of Salt Lake to rapid development. But its influence will be much more than local, for it makes a long desired connection between the intermountain region and the coast of southern California. It would be difficult to imagine a more perfect interchange of natural products than that which is now feasible by means of this line. The semitropical regions of the south need the variety of things which are produced in the higher altitudes and colder climate of the north. There will also be a natural interchange of people between the two localities, the northerners seeking California in winter and the southerners availing themselves of the mountain resorts of Utah, Colorado, Idaho, and Montana, in the summer season.

Enormous improvements in the Central Pacific and Union Pacific system, including the construction of the celebrated Lucin Cut-off across Great Salt Lake, occurred almost simultaneously with the construction of Senator Clark's road to Southern California.

CHAPTER IV

THE CRUDE STRENGTH OF IDAHO

Two travellers crossing Idaho on the same day, one by the Northern Pacific and the other by the Oregon Short Line, would receive quite opposite impressions of the country. The one who had traversed its northern end would think of Idaho as a land of dense forests mirrored in the surfaces of beautiful lakes, of narrow valleys presenting but meagre scope for agriculture, of abundant verdure, and of Alpine scenery. These conditions suggest nothing except the lumber-camp, the mine, and the stock-range.

The traveller who crossed the southern part of the State, on the other hand, would receive the impression of an arid land, with wide stretches of valley and plain covered with wild grasses or sage-brush, alternating with curious formations of rock and lava. This traveller would understand how a large agricultural population may be maintained by turning the abundant water of the streams upon the rich valley soils. Both of these impressions of the resources of the great inland State of the Pacific Northwest would be true, but either of them taken alone, as is often done by travellers, would be quite inadequate. The fact is that Idaho, perhaps even more than other localities in the Far West, presents a

THE CONQUEST OF ARID AMERICA

marvellous diversity of soil, of climate, and of natural endowments. This diversity must necessarily mark its future industrial life and be reflected in the social side of its civilization.

The first important item in the material wealth of Idaho is its water supply. Along its eastern boundary nature has piled up towering mountain-ranges, which receive an enormous snowfall. These mountains are covered with forests, ranking among the most magnificent in the world, which treasure the snow within their sombre depths until the warm weather gradually sends it down to streams which reach out through hundreds of miles of lower valleys. The great river of Idaho is the Snake, which deserves a better name in spite of its tortuous meanderings. This is the largest tributary of the Columbia, and drains a vast water-shed, beginning in the Yellowstone Park of Wyoming and including all of southern and much of western Idaho with eastern Oregon and Washington. Along its course it receives numerous minor streams which drain interior mountain systems. The Snake is nearly one thousand miles long and so deep that in some places soundings of two hundred and forty feet have failed to find the bottom. While incalculably valuable for irrigation, this is by no means its only utility. It is navigable for one hundred and fifty miles above its junction with Clark's Fork in the northern part of the State, and may sometime furnish a water route to the Pacific Ocean through the Columbia. It also has immense possibilities in the way of power, which are now being rapidly developed for the production of electricity, and to furnish light, heat, and power to a number of growing towns.

THE CRUDE STRENGTH OF IDAHO

The most marvellous of these water-powers is furnished by the Great Shoshone Falls, in the south-central portion of the State. Here is a waterfall scarcely inferior in power and grandeur to Niagara, and, like the latter, destined to be an important economic factor in the region within its reach. The abundant water supply is by no means limited to the splendid valleys in the southern part of the State. It is found in hundreds of mountain streams throughout the central portion, and, in the narrow district which tapers northward to the British Columbia line, is so marked a feature of the landscape as to impress the most casual observer. Here Clark's Fork of the Columbia, draining the Bitter Root mountains in western Montana, is a stream of noble proportions. Lakes Pend Oreille and Cœur d'Alene are among the most notable of inland waters, both in beauty and extent. But these northern streams will not be used extensively for irrigation, as there is considerable rainfall and comparatively little agricultural land. They are valuable, however, in connection with mining, lumbering, and water-power.

The forest area of Idaho includes seven million acres, and the principal native trees are fir, spruce of the white, red, and black varieties, scrub oak, yellow and white pine, mountain mahogany, juniper, tamarack, birch, cottonwood, alder, and willow. Some of the large forest regions, notably that of the Pend d'Oreille in the north, are almost unexplored, and constitute the wildest parts of the continent. Naturally, a country so well wooded and watered is the home of fish and game of the rarest kinds. The mineral resources are well distributed and diversified to the last degree. The annual

THE CONQUEST OF ARID AMERICA

output of precious metals now amounts to over twenty-one millions. Gold was discovered in 1860 and mining is still the industry of first importance. The popular excitement over the Thunder Mountain mines in 1901 is still remembered. The chief products are gold, silver, copper, and lead, and there are reduction works for the treatment of these ores employing a large population. Base metals, precious stones, and building materials, including fine marble, exist in abundance.

Idaho lies wholly in the temperate zone, yet its climate presents a great variety of features. In the larger portion of its territory, which consists of mountains and elevated valleys, the winter is a season of considerable severity. Here the thermometer registers far below zero, though the days are rendered comfortable by a dry atmosphere and abundant sunshine.

Southern and western Idaho differ materially from the eastern, central, and northern districts. The altitude ranges from two thousand to four thousand feet. While there are occasional instances of a temperature twelve degrees below zero, the winter in this part of the State is really short and mild, being influenced by the Chinook winds, which make their way from the Pacific over a distance of five hundred miles. Spring opens early and is apt to be windy. The summer temperature is high, though the nights are invariably cool. The almost complete absence of rain between spring and late autumn makes the best conditions for irrigation, though it also involves dry roads and clouds of dust when the wind is high.

Of the healthfulness of Idaho, it is enough to say that

THE CRUDE STRENGTH OF IDAHO

it shows the smallest percentage of deaths of any State or Territory in the Union. This is not only the official record of the population as a whole, but it is the showing of the Army statistics, which furnish a better test, because the conditions of life in that service are remarkably even throughout the country.

Idaho has been called "the Baby State," and in certain respects it seems like a lusty infant, or perhaps now more like an adolescent youth, whose character is just forming. Already there have been four periods in its history. The first was that of the explorer, when Lewis and Clark, and later Bonneville, came to look over the country and report upon its possibilities. The second was that of the trapper, when the Hudson Bay Company established its supremacy after a brief struggle with American hunters. The third was that of the missionary, who as early as 1836 established the first feeble beginnings of civilization, and then pushed westward for the conquest of Oregon. The fourth was that of the miner, who gained a lasting foothold among the mountains and along the streams. The fifth era is now in progress, and has been, after a fashion, since the early sixties. This is the era of agricultural settlement and of town-building. It amounted to little until the railroads were built across the northern and southern extremities of the State, and until enterprise was attracted by the possibilities of irrigation. Agriculture is still second in importance to mining, and is hard pressed by stock raising; but it is a vigorous and growing interest.

With an area larger than New York and Maine combined, and almost equal to the combined area of Pennsyl-

THE CONQUEST OF ARID AMERICA

vania and Ohio, Idaho had a population of only 161,772 by the last census, although it had increased ninety-one per cent. between 1890 and 1900, and is still rapidly growing. The State has twenty-one thousand farms, covering five million acres, and worth two hundred million dollars. The astonishing fertility of the soil is shown by the fact that the average value of the yield per acre of agricultural products in Idaho is nearly double that for the whole United States. The grand prize for the best display of agricultural products at the Louisiana Purchase Exposition in 1904, was awarded to Idaho, besides seventeen gold medals, fourteen silver medals, and twenty-two bronze medals, to individual exhibitors from the State. The chief agricultural product is alfalfa, which yields from five to nine tons per acre on irrigated land. Red clover is also a favorite crop, and wheat, oats, barley, and potatoes are largely cultivated and yield good returns. The cultivation of sugar beets has made great progress since 1903, when the first large factory was erected at Idaho Falls. Since then, three more factories have been built, and the industry is now firmly established and will unquestionably continue to grow in volume and importance.

The horticultural interests of Idaho promise soon to take a place of leading importance. Apples, pears, and prunes are most raised, in the order named. Six million fruit trees, half in bearing, and half of them apples, cover fifty thousand acres of orchards. Peaches, plums, apricots, nectarines, cherries, and grapes also flourish. The solid silver loving-cup given by Senator William A. Clark, to be awarded by the Eleventh National Irrigation

THE CRUDE STRENGTH OF IDAHO

Congress at Ogden, Utah, in 1903, for the best display of irrigation-grown fruits, was won by an exhibit from the New Plymouth Colony, in the Payette Valley, Idaho. At the Louisiana Purchase Exposition, the gold medal for the best display of fruits was won by Idaho, besides many awards to individual exhibitors. The prunes of Idaho are famous, and won the first prize at the Columbian Exposition, Chicago, in 1893. The apple crop is of excellent quality and very profitable. Eastern experts noted with surprise at Chicago that the Yellow Newton Pippins were "twice as large as the same apple grown in the Hudson River Valley" of New York.

The greatest irrigation development in Idaho heretofore has occurred in the upper Snake River Valley. The sixteen southern counties have a total of 3577 miles of canals and ditches, costing nearly ten million dollars and covering 2,108,095 acres, of which 835,115 are cultivated. The large canals are owned by private companies and usually represent Eastern capital.

The most notable colony yet made on the irrigated lands of Idaho is that of New Plymouth, in the Payette Valley, twelve miles from the town of Payette. This colony, organized in the spring of 1895 by William E. Smythe and Benjamin P. Shawhan, was intended to represent a high social and industrial ideal. The initial work of enlisting settlers and public interest for the undertaking was done in Boston, with the aid of Dr. Edward Everett Hale and other prominent men, but most of the actual colonists were from Chicago and the Middle West. The pioneers of New Plymouth, who represented a rather unusual quality of settlers, were drawn principally from urban business and professional life, yet entered enthusiastically and successfully upon the work of making homes on sage-brush land twelve miles from a railroad, in a remote and undeveloped part of the West.

The Plymouth industrial programme aimed at complete economic independence of the people by the simple method of producing the variety of things consumed, on small, diversified farms; of having surplus products, principally fruit, for sale in home and Eastern markets; and by combining the capital of the settlers, by incorporation of a stock-company, to own and develop the town-site, and to erect and operate simple industries required in connection with the products of the soil. On the social side, the plan aimed to give these farmers the best advantages of town life, or at least of neighborhood association. This was accomplished by assembling the houses in a central village, laid out, in accordance with a beautiful plan, with residences grouped on an outside circle touching the farms at all points. This plan brought the settlers close together on acre-lots—"home-acres"—thus preventing isolation, and giving them the benefit of school, church, post-office, store, library, and entertainments.

The Plymouth settlers have been contented and prosperous from the first, and have had less than the usual share of early trials and disappointments. They testify that the social advantages of the colony plan, as compared with the drawbacks of individual and isolated settlement, are alone sufficient to warrant its use. Availing themselves of a favorable opportunity, they acquired the irrigation system and other valuable property by

THE CRUDE STRENGTH OF IDAHO

purchase from the Eastern bondholders, on terms which went far to enrich them as a community.

A great stimulus has recently been given to the reclamation of arid lands, especially in the lower Snake River Valley, by the operation of the "Carey Act." Already a third of a million acres of public land have been withdrawn from settlement and turned over to the State Land Board for reclamation under this law. The Board has entered into agreements with several companies for the construction of irrigation works, to redeem tracts of this land ranging from six thousand to two hundred and forty-four thousand acres.

Under the liberal provisions of this law, the Nation gives the land to the State on condition that it be reclaimed, and the State sells it to actual settlers at the nominal price of fifty cents per acre, in tracts limited to one hundred and sixty acres for each settler, on conditions of actual residence and improvement and that the settlers shall repay the construction company the cost of the works. This cost is agreed upon in advance by the State Land Board and the construction companies, and ranges from ten to twenty-five dollars per acre. Easy terms of payment are arranged, sometimes as much as ten years time being allowed.

The most important project under the "Carey Act" is that of the Twin Falls Land & Water Company, said to be the largest private irrigation enterprise in America. It will reclaim two hundred and forty-four thousand acres of land lying south of the Snake River and thirty thousand acres on its north side. The headworks are at Twin Falls, at the head of the Grand Canyon of the

THE CONQUEST OF ARID AMERICA

Snake River, where the new town of Milner has sprung up. The main canal has a length of nearly seventy miles, and ends at the Canyon of the Salmon River. The "Minidoka Project," now under construction by the National Reclamation Service and elsewhere described in this book, is situated fifty miles up stream from Twin Falls. Work on the Twin Falls project was begun in March, 1903, and water was turned into the canals just two years later. The land has been rapidly taken up and crops are already growing under the canals. The immense dams at the headworks are 1900 feet long, from 60 to 76 feet high, and raise the normal level of the river forty-nine feet. The gates and spillways, constructed in the rocky islands of the river, are marvels of engineering skill. The main canal is 80 feet wide on the bottom, 112 feet at the surface of the water, and carries a stream 10 feet deep. There will be nearly a thousand miles of canals and laterals in the system. Gasoline launches are to be put upon the main canal to carry passengers and freight. The total cost of the works will exceed two million dollars.

The prices of land in Idaho naturally vary greatly with location and conditions. Unimproved and unirrigated lands can be had at nominal prices. Improved lands without water bring from five to twenty dollars per acre. Lands reclaimed under the "Carey Act" cost, with a permanent water right, from ten to twenty-five dollars. Where the reclamation work is being done by the Government, the cost per acre is estimated at twenty-six dollars. Unimproved lands under existing private canals bring from twenty to fifty dollars per acre, and improved irrigated lands are worth from forty to two hundred dollars or more. While the chief agricultural and horticultural districts of the State lie along the Snake River and its important tributaries, the mountains of central Idaho are full of picturesque, well-watered valleys, in which settlement has been made for more than a generation. In these high altitudes, however, production is limited to hardy crops and runs largely to hay and grain, which find a market in the surrounding mining and lumbering camps and stock ranches. The Nez Percé Indian Reservation is a fertile and promising country, though the Indians have been located in severalty on some of the most desirable lands, which would otherwise be open to settlers. A considerable locality in the northern part of the State, known as "the Palouse Country," is farmed in grain without irrigation. The same is true of the Cammas Prairie, in one of the central counties. But Idaho is substantially an arid region and its characteristic institutions are growing up where irrigation has been supplied. The ultimate development of its diversified resources will give it a many-sided economic life.

Each of the early sources of Idaho's growth left its driftwood along the slender stream of the State's development. The "old-timer" is an influential element in its citizenship. Later comers, perhaps forgetting the distance which has been covered since the days of the primeval wilderness, and, in their impatience for progress, inclined to belittle the hardy heroism which made it possible, sometimes complain that the "old-timers" are content to live in the memory of "the early days" while contributing little, either of enthusiasm or capital, to further development. The obvious truth is that different classes of people are required for different classes of work. If the men who filled the rôle of pioneers are not well suited by tastes and temperament to solve the problems involved in the evolution of a complex industrial life, it is doubtless equally true that the element which enters enthusiastically and intelligently upon this later work would not have dealt as successfully with the harsher conditions of thirty-five years ago. It is true, however, that there are two well-defined classes in the citizenship of the Northwest, and that they represent different ways of thinking. The steady growth of population in Idaho has already given the supremacy to those who are trying to put the farm in place of the desert, to develop the best methods of fruit-culture, to bring the irrigation system under rigid public supervision, and to establish the highest standards in political and social life.

CHAPTER V

THE GIANT WASHINGTON

WASHINGTON is one of the big States of the West—big in resources and potentialities—and big in its yearning for the surplus man. In thinking of its future, it is difficult for the human imagination to set bounds to its development. It is blest in the possession of everything which goes to the making of a complete industrial, social, and political organism. Anywhere else than in the United States it would be considered amply sufficient for the sustenance of a nation. Its resources are wonderful alike in richness and in variety. Even its climate presents as many diversities as the entire Atlantic Coast Line, from Eastport to Savannah. If there be such a thing as a psychological influence arising from location and from proximity to the seat of events, and affecting the destiny and character of communities, then that influence is operative in Washington.

The State is fortunate not only in having a strategic location and marvellous natural resources, but in being able to command the interest and support of the large capital which is essential to enable it to take fullest advantage of these facts. It has been the beneficiary of the most daring and aggressive railroad enterprise from the beginning of its development. This has enabled its principal cities promptly to become the commercial cen-

THE CONQUEST OF ARID AMERICA

tres of the rich agricultural, mining, and timber districts which lie behind them. The sudden birth of Alaska as a new and powerful economic quantity found Washington able to respond instantly and adequately to the new demands. The opening of China and other Oriental markets to American products again found Washington able to rise to its opportunity, this time with a fleet of the largest freight steamers in the world. If mere possession of rare natural advantages made cities, with thrifty, growing surroundings, then many localities now silent and stagnant would long since have assumed importance in relation to the commerce of the world. But natural advantages are not alone sufficient—they must be quickened into life and productiveness in order to count largely in the making of wealth and the support of communities. Washington owes very much to the men of daring enterprise and enormous financial resources who stretched lines of steel from Puget Sound to the Great Lakes and who, refusing to stop at the shore-line, utilized the free highway of the sea that the markets of Alaska and the Orient might be opened to the trade of this young commonwealth.

In the decade covered by the last census, Washington increased its population forty-five per cent., notwithstanding the fact that the larger portion of the period fell in hard times, when some mushroom cities and towns actually declined with the recession of the boom. Its present total of a little more than half a million represents a mere squad of people compared with the vast army who will ultimately be sustained in the varied industrial life of the State. That is to say, Washington is in its infancy

THE GIANT WASHINGTON

and must inevitably enjoy constant growth, which will accelerate as time goes on. This will be due in part to its own natural wealth and in part to its fortunate location, which makes it the commercial partner of Alaska, of the Orient, and of Northwest Canada in what promises to be the most stirring era of material and commercial development which the world has ever seen.

Puget Sound is one of Washington's priceless possessions. It is not merely a harbor, but a beautiful inland sea, sometimes, and not inappropriately, referred to as "the American Mediterranean." Coal, iron, and timber are abundant, and the shores of the Sound must in time be lined with a great variety of the most substantial industries. Those already well established include the manufacture of timber and lumber products, of flour and grist mill products, the canning and preserving of fish, copper smelting and refining, the manufacture of iron and steel products, and of paper and wood pulp. Manufacture is, however, only in its initial stages. Its future expansion, with the attendant growth in mercantile, banking, and professional lines, will absorb many surplus men. But here, as in all the other new States of the Far West, the conquest of the soil will make the largest demand for future recruits.

Like Oregon, Washington presents some sharp contrasts in climatic conditions. On the coast the rainfall is excessive, resulting in a dense growth of natural vegetation, including valuable forests. The extreme eastern portion of the State, while receiving much less rain, is also able to produce large and fairly regular crops without artificial moisture. But the central valleys, lying

THE CONQUEST OF ARID AMERICA

between the Cascade and Bitter Root Mountains, belong distinctly to the arid region. Here irrigation produces wonderful results. The Yakima Valley is the leading example of irrigation development in Washington. It is one of the few places in the West where water is relatively more abundant than land and, consequently, where there need be no limitation upon growth by reason of lack of moisture. The Columbia River flows in a deeply eroded channel and far below the level of fertile land which could be made immensely productive with irrigation. It will be the labor of some future generation to utilize such opportunities when, with growing density of population and increase of land value, it will be economically profitable to do many things which are now impracticable. In the meantime, streams which may be diverted readily and cheaply upon the rich lands of the valleys through which they flow, abundantly demonstrate the economic significance of irrigation in Washington. And chief of such streams is the Yakima.

All students of Western resources agree that the irrigated farms of central and eastern Washington are among the most prosperous to be found in the West. They are such, to begin with, because of the large and regular water supply which has already been noted. The soil, largely of volcanic ash, is rich, deep, and easily worked. The climate and range of products are most favorable, while the home markets are peculiarly attractive. These latter include not merely the cities of Puget Sound, with their growing export trade, but the great mining camps of British Columbia, Montana, and northern Idaho. The vegetables and small fruits of the

THE GIANT WASHINGTON

Yakima country are only about three weeks later than those of California. These early products command ready sale at fancy prices in the mining districts, which are reached by direct lines of transportation. From the heart of Yakima Valley to Rossland, British Columbia, is but eighteen hours; to the Coeur d'Alêne of Idaho, but twenty hours; to the great mining towns of Montana, but twenty-four hours. This matter of early products, good transportation facilities, and large home markets is a consideration of highest importance to settlers. It enables them to reap early and large reward from their labor, since crops which may be harvested the very first season after planting sell at good prices under these favorable conditions. It is sometimes possible for settlers to pay for their land in a single year. In the absence of any one of the three factors which have been mentioned, this could not be done.

The growth of population in the irrigated districts has been large and rapid during the past few years. This has been materially influenced by the Alaska boom, a circumstance which supplies an admirable illustration of the peculiar advantages of Washington. In his speech at Seattle, President Roosevelt predicted that one of the largest and most prosperous American States would grow up in Alaska. Washington is the nearest agricultural field, under the American flag, to that great northern Territory which can never produce most of the food products which it must so largely consume. This fact has enormous significance in relation to the future prosperity of the irrigated valleys of the Columbia and its tributaries. Their orchards will furnish dried fruits, their gardens vegetables, their alfalfa fields cattle, pigs, and their products, to be shipped in large and ever-increasing quantities to the inhabitants of the Frozen North. The peaches, prunes, and apples of this locality are especially fine and well adapted, when dried or preserved, to the demands of the export trade. Dairying and poultry-raising are also profitably carried on here. For all these reasons, it is perfectly safe to predict that many people will find attractive opportunities to participate in the making of irrigated Washington.

The climate of this extreme northwestern State is frequently misunderstood at the East. As in the case of California and Oregon, so also in Washington, we find that northern latitude has little or no significance in relation to summer and winter temperature. Indeed, Washington rejoices in the soubriquet, "The Evergreen State." When one crosses the continent by the most northern transcontinental route, through the rigors of the Canadian winter, the influence of the Japanese current and its famous chinook winds is felt the moment one begins to pass down the western slope of the Rockies, into British Columbia. And in January or February one need not be surprised to look upon green lawns in Victoria and Vancouver. The Puget Sound region must plead guilty to one climatic indiscretion—it drizzles persistently, month in and month out, during the winter season. That is, it drizzles when it does not rain "right smart." Gray skies are characteristic of the North Pacific coastline. On the other hand, nothing could be more delightful than the many clear and beautiful days which the Puget Sound region does enjoy in the course

THE GIANT WASHINGTON

of a year. Then the waters of the inland sea mirror the blue of the sky, and the great white dome of Mount Rainier (if you happen to be in Seattle; otherwise, Mount Tacoma), with the lesser heights of the Olympic Mountains, presents a picture of grandeur which may never be forgotten.

The larger portion of the State, lying east of the Cascade Range, presents very different conditions. There are places on the Columbia River, in the very heart of the State, where the rainfall is lighter than in any other part of the United States, with the exception of the southwestern deserts, in Arizona and California. Such places enjoy an extraordinary number of cloudless days and may very properly be classified with the Land of Sunshine. In this portion of the State, the altitude is much higher than in the Sound region and the seasons more sharply differentiated. Along the Columbia plowing may be done almost continuously, while at the higher elevations it is usually suspended from the middle of November to the middle of February.

Eastern Washington produces grain without irrigation and is a famous grazing country. This is due probably as much to the remarkable quality of the soil as to the rainfall, which is not extensive. The Great Bend region is noted for its wheat, while the " Horse Heaven " country enjoys a renown in connection with the grazing industry which is sufficiently indicated by its name.

The principal towns in the irrigated districts of Washington are North Yakima, Prosser, Wenatchee, Ellensburg, and Walla Walla. Kennewick and Pasco, on opposite sides of the Columbia, near its confluence with the

THE CONQUEST OF ARID AMERICA

Yakima River, are growing and destined to be of much more importance in the future. Spokane is the attractive metropolis of eastern Washington. It has lived through the rise and fall of the inevitable boom and now rests upon most substantial foundations. These are the industries of mining, lumbering, stock-raising, and grain-growing.

CHAPTER VI

OREGON IN TRANSITION

OREGON is one of the finest of all Western States and one which is not appreciated as it deserves to be. Its population, according to the census of 1900, is only 4.4 persons to each square mile. By a curious coincidence, it gained almost exactly one person for each square mile of its area during the last ten years, a rate of growth which would require many centuries to enable it to approach anything like the limit of its capacity for the support of population. For those who know Oregon, it is not difficult to understand the backwardness of its development.

It is only about twenty years since the transcontinental railroad reached Portland, and the chief growth of the State has been realized since that event. Existing transportation facilities are entirely inadequate. They permit Oregon to live; they do not enable it to develop its resources as they ought to be developed. It is as though Boston had one railroad up the coast from New York and another transcontinental line from the West,— "only this and nothing more." Instead, the group of six small New England States is gridironed with steam and electric railroads, penetrating every district of the slightest interest or value and bringing the blood of trade through a thousand arteries to the commercial heart at Boston. With such facilities, it is possible to develop

THE CONQUEST OF ARID AMERICA

natural resources to the limit and to maintain a dense population.

Oregon has two railroads, one running north and south, the other east and west. True, Portland also enjoys connections with the three northern railways which send spurs down from Puget Sound; but while these serve the needs of through traffic to some extent, they do not assist the internal development of the great commonwealth. They do not open up the splendid interior valleys to settlement, nor bring the products of the mines, farms, orchards, and ranges to the seaboard. They do not break the isolation of remote districts which have all the raw materials for an active and prosperous economic life except population and which cannot hope to have population until transportation facilities are provided. Thus the passing years do but little to widen the foundation of Oregon's industrial life, notwithstanding the fact that priceless resources stand ready and waiting.

Oregon fell an early and easy prey to the spirit of bonanza farming. Its virgin soil bore extraordinary crops of small grain, sometimes yielding three or four crops as the reward of a single sowing. Land was cheap and easily acquired. There was plenty of room for big things, including big farms. Hence, the grain industry took firm possession of the beautiful Willamette Valley, of the Palouse country, and of other favored localities. The single-crop idea was widely prevalent. Here, as always and everywhere, it ended in the impoverishment alike of the soil and of the farmers. With the fall of prices and the pressure of indebtedness, this most extensive of Oregon industries has suffered severely until

OREGON IN TRANSITION

at last a reaction appears to have come in favor of smaller farms and more diversified production.

Lumbering, mining, and manufacturing have been considerably developed and are capable of enormous expansion in the future. Manufacturing is hampered, however, by lack of transportation facilities and of cheap fuel, and by the limitations of the market. It would seem that the great opportunities of Oregon in the future lie in two directions. First, in the way of internal development, chiefly by the irrigation of its fertile soil and its subdivision into a multitude of small estates; second, in the growth of its export trade to Alaska and the Orient. As to the first, it must rely in part upon the National Government and in part upon its own energies, public and private. The second opportunity is in control of world-wide forces now apparently in active operation.

The moist coastline of Oregon has given it wide reputation as a land of excessive rainfall. It may, therefore, seem surprising to speak of this State as needing irrigation. The fact is that a portion of it receives the heaviest rainfall in the United States. This is on the western slope of the mountains which run down close to the coast. The western third of the State is sharply divided from the eastern two-thirds, so far as its climatic character is concerned, by the lofty Cascade Range. This intercepts the moisture-laden clouds and winds from the Pacific, condensing them into snow and leaving the larger part of Oregon arid or semi-arid. Thus it happens that no other influence would go so far to quicken and to broaden the economic life of Oregon as the intelligent application of a great irrigation policy. It would effect a sweeping revolution in its entire industrial character. It would arrest the downward tendency of agriculture by stopping the impoverishment of the soil, by abolishing the evils of the single crop, and by fostering that spirit of co-operation without which the farmers must always be exploited for the benefit of those who furnish their supplies and handle their products. On the social side, it would give to the future population the innumerable advantages which come with density of settlement and which are wholly unattainable where isolation is the prevailing condition.

Considered from every standpoint, irrigation is the golden key which may alone unlock the doors of civilization to wide districts in Oregon, particularly east of the Cascade Mountains. The logic of these facts is gradually coming to be realized and accepted. New tendencies are beginning to show in the life of the State. If not yet very powerful, they are growing. There can be little doubt that by the first or second quarter of the present century, Oregon will present an entirely different picture, from the industrial standpoint, than it does today.

The two railroads which traverse the State give the traveller a fair idea of its character as a whole. Entering from the south and journeying north to Portland, one rides for hours through the charming valleys of the Rogue and Willamette rivers in the midst of beautiful scenery and within sight of pine forests and fertile farms. Of late years the orchards, mostly of apples, pears, and prunes, have notably increased, while in the vicinity of many thrifty towns there are grateful signs of the growth of small homes. The climate of this western valley, which lies directly north of the Sacramento, is not very different from that of California, though the variation in temperature is just sufficient to destroy the California characteristic in certain lines of production. Oranges and lemons are not grown commercially. Many people think the climate superior to that of California because it offers a little more variety, while retaining the mild quality which is so attractive and healthful. The conditions are extremely favorable for the development of the finest sort of colonies which, with irrigation as a foundation, might be almost wholly self-sustaining and have a large surplus product to dispose of in the home market. This is true for the reason that Oregon as a whole by no means produces all it consumes, and that farmers engaged in raising grain are themselves large patrons of those who take varied products from the soil.

Portland is a thriving commercial and manufacturing city, as well as a financial, social, and political centre. Its position on the Willamette near its confluence with the Columbia, though nearly a hundred miles inland, enables it to enjoy a considerable coastwise and foreign trade. It is one of the wealthiest and most substantial of American towns and, both in business and residence districts, presents a quality of solidity that is unusual in new countries. While its position as the foremost city of the Northwest is, perhaps, menaced by the phenomenal growth of Seattle, it will always be the metropolis of a vast region.

The railroad running east from Portland into south-

THE CONQUEST OF ARID AMERICA

ern Idaho crosses the Cascade Mountains and follows the Columbia River for a considerable distance. It penetrates the plains and rolling hills, which are chiefly devoted to wheat-farming, and passes on through the mining regions in the Blue Mountains and the sage-brush deserts which mark the most easterly counties of the State. Here settlement is extremely sparse and, in most localities, development has not proceeded beyond the frontier stage. It is a prosperous stock country, but large portions of it are capable of better things. To quote F. H. Newell:

"It may be pictured as a series of broad plains and mesas, covered with lava of various ages, from that outpoured recently to the ancient flows whose surface has largely changed into soil. This supports a dense growth of sage-brush, and also juniper near the mountains, these being intermingled with forage plants. The vegetation becomes sparse out on the broad valleys, but nearly everywhere furnishes good grazing."

Here the land is chiefly the property of the Government. Large areas are susceptible of irrigation and of serving as the foundation of a very desirable class of homesteads. The water supply is quite abundant, as the Columbia receives several large tributaries from the south which might readily be turned upon the soil. Among the valleys where large opportunities for reclamation may be found, chiefly by means of storage, are those watered by the Snake River, the Malheur, the Des Chutes, the John Day, and Willow and Bully Creeks. There are also many large natural lakes in eastern Oregon which might be drawn upon for irrigation. There are large possibilities in connection with the development of underground supplies by means of artesian wells.

The altitude of this district ranges from three thousand to four thousand feet, though the mountains reach eight thousand feet or more. The climate is entirely different from that of western Oregon. There is a wide range of temperature during the course of the year—a good deal of honest cold in the winter and of honest heat in the summer. The winter also brings snow and ice, though these do not remain long at a time. It is the familiar climate of the temperate zone, agreeably tempered by the prevailing aridity of the Far West.

Taken as a whole, Oregon is capable of furnishing work and homes for large numbers of men and women. Under wise laws governing its development, it could absorb hundreds of thousands into its agricultural life and thousands more in the use of other natural resources, such as the forests, the mines and quarries, the grazing lands and water powers. These developments would necessitate more thousands in the various employments of numerous towns. To put it in a word, Oregon is a great State, now only in the infancy of its development, waiting to be used when men shall have need of its resources and shall be wise enough to fit their laws and institutions to the conditions it offers them.

Oregon is to-day in a state of transition. It has been known as a land of excessive rainfall, and it is now to be celebrated because of its triumph over the aridity which prevails throughout much the larger portion of its domain. It has had large farms and the single crop; under irrigation systems which are being rapidly extended, both by private and public enterprise, it is to have small farms and diversified productions. Its development has been notably slow; it now promises to go forward by strides and bounds in response to new influences which are rapidly rising in its economic life. The success of the Exposition in commemoration of the Lewis and Clark centennial will greatly strengthen these new tendencies and Oregon may be expected to give a good account of itself when the next national census is taken.

CHAPTER VII

THE RISING STATE OF NEVADA

NEVADA, after a period of stagnation and decline, is moving along the upward path with steady strides and stands well to the front among States which are conspicuously prosperous.

No mining camps are attracting wider attention than Tonopah, Goldfield, and Bullfrog. No new agricultural district is more prominently in the eye of the homeseeker than Carson Valley, watered by the first government canal to reach completion. No railroad developments now in progress promise more revolutionary results in opening rich, but hitherto idle, natural resources to human conquest, than the "Clark Road," which traverses the neglected empire of southern Nevada, the Western Pacific, which is to cross the State from east to west, and the lines which have been extended into the new and flourishing mining camps near the southwestern border. And few indeed are the towns which show a stronger pulse-beat than Reno, the commercial capital of the State.

No division of the Union has been so persistently and grossly misunderstood as the big sage-brush commonwealth which lies between Utah and California—two States of unusual human interest. The popular impression of Nevada has been largely created by those whose opinion of its scenery and resources is based on their experience of a railroad flight across its wide expanse. They glance impatiently out of the car window, inhale some alkali dust, and then denounce the region as "only fit to hold the earth together." If they happen to be literary artists, they vent their disgust in some such striking phrases as these, employed by a popular writer in a recent novel:

"For beauty and promise, Nevada is a name among names. Nevada! Pronounce the word aloud. Does it not evoke mountains and clear air, heights of untrodden snow and valleys aromatic with the pine and musical with falling waters? Nevada! But the name is all. Abomination of desolation presides over nine-tenths of the place. The sun beats down on a roof of zinc, fierce and dull. Not a drop of water to a mile of sand. The mean ash-dump landscape stretches on from nowhere to nowhere, a spot of mange. No portion of the earth is more lacquered with paltry, unimportant ugliness."

What a difference in human souls! The man who sees a "spot of mange" in God's handiwork only reflects the spot of mange within himself, and shows how his own intelligence is "lacquered with paltry, unimportant ugliness." John C. Van Dyke looks upon the same scenes and then writes, in that classic, "The Desert:"

"Not in vain these wastes of sand. And this time not because they develop character in desert life, but simply because they are beautiful in themselves and good to look upon whether they be life or death. In sublimity—the superlative degree of beauty—what land can equal the desert with its wide plains, its grim mountains, and its expanding canopy of sky! You shall never see elsewhere as here the dome, the pinnacle, the minaret fretted with golden fire at sunrise and sunset; you

BUILDING A GOVERNMENT CANAL IN NEVADA.

THE RISING STATE OF NEVADA

shall never see elsewhere as here the sunset valleys swimming in a pink and lilac haze, the great mesas and plateaus fading into blue distance, the gorges and canyons banked full of purple shadow. Never again shall you see such light and air and color; never such opaline mirage, such rosy dawn, such fiery twilight. . . . Look out from the mountain's edge once more. A dusk is gathering on the desert's face, and over the eastern horizon the purple shadow of the world is reaching up to the sky. The light is fading out. Plain and mesa are blurring into unknown distances, and mountain-ranges are looming dimly into unknown heights. Warm drifts of lilac-blue are drawn like mists across the valleys; the yellow sands have shifted into a pallid gray. The glory of the wilderness has gone down with the sun. Mystery—that haunting sense of the unknown—is all that remains."

The difference between these two authors is only a difference in development. The one beholds a sealed book; the other understands. Nevada is typical of the whole desert region between the Rockies and the Western Ocean. To those who cannot comprehend its strange ensemble it is undeniably ugly, but to those who can comprehend, it is a land stamped with a beauty full of endless surprises. These latter are not necessarily cultured Van Dykes. They may be men who have never studied art or even read a book. Many a Piute Indian has looked upon the deserts and mountains of Nevada with a comprehension utterly denied to the novelist who beholds nothing in the scene except a " mean ash-dump landscape."

Even the fleeting railroad tourist might correct his superficial impression of Nevada's worthlessness by getting out of the car occasionally. Let him step off for a few moments to enjoy the cool fragrance of the little oasis at Humboldt, to walk within the shade of its trees and hear the music of its waters. The little patch of green which a hillside spring has spoken into being is a sample of what millions of desert acres will become. Farther on, the west-bound traveller catches a twilight glimpse of the thriving farms of Lovelock or the green Truckee meadows. But the larger examples of irrigation lie off the beaten path. Such an instance is the Carson Valley, hidden between the sheltering shoulders of the Sierras. To appreciate the possibilities of the region, the critic should visit that valley in the perfect Nevada springtime and look upon its farms, its homes, and its villages. There he would behold a memorable picture of thrift, of beauty, and of peace, from the white blossoms in the dooryards to the white summits of the mountains, and there he might read the true prophecy of Nevada's future.

Nevada farmers are very prosperous on the average, taking one year with another, and probably much more so than the farmers in more pretentious localities. For the most part, they were poor when they came and have grown steadily better off. The climate is perfectly adapted to the production of all the cereals and hardy fruits. The wheat is perfect, with a full, rich kernel and a clean, golden straw, free from smut and rust. It has taken prizes at all the great expositions. With a variety of soil, on the different slopes of hillside, plain, and valley, there are conditions to meet almost every requirement in an agricultural way within the limitations of climate. It seems absurd to explain that Nevada does not produce oranges, yet the question is sometimes asked

WHERE THE GATES WERE LIFTED.—First Government Project, Dedicated June 17, 1905, Truckee River, Nevada. (On line of the Southern Pacific.)

THE RISING STATE OF NEVADA

by those who only know that Nevada is the next-door neighbor of California. Speaking broadly, Nevada is an elevated plateau in the Great Basin enclosed by the Wasatch Range on the east and the Sierra on the west, having an average altitude of about four thousand feet. Its climate is that of the north temperate zone. The winter is cold, the summer hot, the springtime marked by showers and high wind, the autumn long and golden. As in other parts of the arid region, the dry air moderates cold and heat, giving man and vegetation the benefits arising from the vigorous qualities of these extremes without the unpleasant effects which are felt in humid districts.

The national irrigation projects in Nevada are described in a later chapter, but it is important to note here the influence which this development must inevitably exert upon the whole social, political, and commercial life of the State. There will be a steady influx of population for many years to come. Farms will be smaller and more intensively cultivated. There will be a corresponding expansion in all lines of business. Social life in the country will lose its frontier characteristics, and political power will gravitate largely into the hands of the hosts of newcomers, drawn from many different parts of the United States. Owing their opportunities to the first great national experiment in the public ownership of utilities essential to industrial development, it would be strange indeed if this new population—the dominant element of the future—does not favor very advanced ideas in politics.

Of all the slanders circulated about Nevada, the story

THE CONQUEST OF ARID AMERICA

that its mineral wealth was exhausted was the most senseless. This slander has been effectually answered by events, for Nevada is to-day the scene of the greatest activity in mining and the centre of attraction for that large public which rushes toward the newest and most glittering camp. Writing of this aspect of its resources, Mr. Robert L. Fulton, one of the most prominent and useful citizens of Nevada, says:

"Nothing could be more tempting to the settler than the chance to find a good mine near home, and there is not a valley in Nevada that does not lie within sight of ranges of hills containing a good percentage of the precious metals. The farmer's boy need not loaf in the saloon or post-office, if he has energy and sense. He can open a prospect and work all winter in a good warm tunnel or a sheltered cut, and, if he finds ore, he can have it hauled to the quartz-mill in the spring, just as his father hauls his wheat in the fall. Mining in Nevada has had a wonderful history, with the record of the finest camp the world has ever seen, where over six hundred million dollars were taken from one rock a mile in length, but the past is but a promise of what is to be when the State is settled and the university has taught our boys how to test the rocks and prospect the hundred thousand square miles of territory that has hardly been scratched as yet."

With few exceptions, deep mining has not been pursued. Only the richer ores near the surface have been utilized, and these by expensive processes and at high cost of transportation. It is interesting to call the roll of the fourteen counties and to observe their mineral possibilities.

Elko, in the extreme northeastern corner of the State, where the railroad traveller enters from Utah, yielded placer-gold to the earliest prospectors of the Great Basin

THE RISING STATE OF NEVADA

and has gold ledges of promising extent and value which are now being carefully explored. Humboldt, central on the northern boundary, presents as great a variety of resources as any district in the United States. Besides silver, it possesses gold, copper, lead, tin, iron, antimony, nickel, cobalt, bismuth, nitre, sulphur, gypsum, borax, soda, and salt. Coarse gold to the value of several millions has been taken from its placer and gravel mines. Gypsum is shipped to San Francisco for fertilizer. Near Lovelock, in this county, are great hills of fine bessemer iron ore, yielding eighty-six per cent. of iron and twelve per cent. of aluminum, with no trace of impurities. Eureka county, in the central part of the State, has many mines in which gold predominates, besides large deposits of magnetic iron ore, of lead, of granite, and of other building stones. Lander, adjoining Eureka on the west, has valuable undeveloped gold deposits and the richest mines of antimony in the world. Of the western counties, Washoe reports recent discoveries of gold, copper, and iron; Douglas, quartz and placer-gold; Lyon, mines which run high in gold, with but little silver; Churchill, gold, copper, and other minerals; while Storey contains the Comstock. Esmeralda, bordering California on the extreme southwest, is very rich in gold-bearing quartz, and is being actively developed. Lincoln and Nye, the two great counties of the south, have gold, copper, lead, antimony, zinc, quicksilver, fire-clay, chalk, soapstone, borax, and alum. In Lincoln there is a deposit of zinc, estimated to be worth several millions, which cannot be worked because of lack of transportation facilities. There are hills of salt, the product of which commands locally but $1 per ton, owing to its inaccessibility, though other localities in the State pay $20 to $40 per ton for a similar product. White Pine County, along the eastern boundary, has extensive gold placers.

Finally, there is a large deposit in Elko County of something which is said never to have been discovered elsewhere—mineral soap, superior in cleansing virtues to any of the manufactured varieties known to the students of modern advertising. As the country was principally occupied by Piute Indians, the deposit remained undisturbed for nameless centuries. But it was exhibited at the World's Fair at Chicago, where, it is to be feared, it added nothing to Nevada's fame. The thing was so palpably and unmistakably the perfection of toilet articles that it over-taxed Eastern credulity and was quietly set down as a larger piece of mendacity than of soap.

Standing on the height above the roaring Truckee at Reno, in the midst of fragrant alfalfa fields and wellfruited orchards, but little imagination is required to behold the Nevada of the future which is now rapidly rising on the Nevada of the past. A big, splendid, American State, blest with the climate in which Englishspeaking man has won nearly all his triumphs, except that its skies are cleared by aridity and its sunshine brightened by altitude, a land full of prosperous little farms, tilled by their owners, mountains pouring out their annual tribute of gold and silver, towns large enough to offer the refinements of modern life yet small enough to escape the awful contrasts between superfluous wealth and hopeless poverty, and a people so economically freed and politically untrammelled that they may make their institutions what they will,—this is the Nevada of the future.

TWO IRRIGATION STATESMEN AND A U. S. ENGINEER.—Engineer Taylor, Builder of the Truckee-Carson Project; Congressman Mondell, Chairman of the Committee on Irrigation; Senator Newlands.
(Courtesy of the Sunset Magazine.)

CHAPTER VIII

THE UNKNOWN LAND OF WYOMING

A SINGLE railroad traverses the length of Wyoming, taking the traveller through that portion of the State possessing the least attractions in the way of scenery and development. As a consequence, thousands of people who have made the transcontinental journey think of this new commonwealth as a barren wilderness of withered grass and stunted sage-brush, with an abundance of rugged mountain views along its southern horizon, but without visible means of support for population save a few cheerless trading towns and grimy coal-mining camps. These tourists find the altitude disagreeably high and the atmosphere generally chilly, if not cold. They behold no cultivated fields, no homes framed in trees and vines; hence do not marvel that the population of this vast State is no larger than that of fourth-class cities in the East.

Spite of this popular prejudice, which may hardly be complained of as unreasonable, Wyoming is a very great State in its natural resources, and must some day sustain a population as large as that of Ohio and Illinois. If its first railroad had penetrated its central or northern counties it would even now be as celebrated and as populous as Colorado. Because of its stores of coal and

THE CONQUEST OF ARID AMERICA

petroleum it is frequently called the "Pennsylvania of the West." Its deposits of both base and precious metals are extensive and widely diffused, though the present output is small, owing to the cost of transportation and the fact that mining capital and enterprise have been attracted elsewhere by the greater fame of other localities. It is well endowed with forests and blessed with the noblest scenery, of which the far-famed grandeurs of the Yellowstone Park furnish the best example. But its greatest resources are those of water and of land. It is estimated that not less than ten million acres of fertile land may be reclaimed by irrigation. Distributed rather evenly through different portions of the State, and surrounded by the wealth of mine, forest, water-power, and natural pastures, this irrigable land will furnish the solid foundation of a great and manifold economic life in future centuries.

The great industry of Wyoming from the time of its first settlement has been stock-raising. Its agriculture has been mostly auxiliary to this. Herds of horses, cattle, and sheep are grazed upon the enormous free pasture or range from spring to autumn, and then fed upon the native or alfalfa hay raised in the irrigated valleys. This industry has been the source of local prosperity and enlisted great sums of eastern and foreign capital. It is a pursuit which does not develop the higher possibilities of the country, either in a material or social way, and so long as its influence strongly dominated the life of the community Wyoming did not furnish an attractive field for settlers. There was a time when prominent men actually deprecated the growth of population, and boldly asserted that brute cattle were

THE UNKNOWN LAND OF WYOMING

more to be welcomed than men, women, and children in that sparsely settled empire. In the last few years, however, the tendency of public thought and political action, consequently of development, has been distinctly away from barbarism and towards civilization.

What is rather grotesquely known as "The Rustlers' War" of 1892 had much to do with the changed conditions. Properly speaking, it was not a war, but a raid, which ended disastrously so far as its immediate purpose was concerned. Individuals and companies owning large herds of horses and cattle had suffered repeatedly from the depredations of thieves or "rustlers." They had often apprehended the culprits and sought by every means in their power to punish them through the courts. But the cases were tried in counties where public sentiment strongly opposed the great cattle-owners. The result was that no jury could be found to convict. After a long and exasperating experience of this kind the large stock interests determined to try a heroic remedy. They fitted out an expedition, consisting mostly of rough characters from Texas, and thoroughly armed it, even a Gatling gun being included in its equipment. The expedition was led by prominent and wealthy citizens and accompanied by a young English lord in search of a new sensation.

A considerable number of "rustlers," who were settlers living in lonely places with small bands of cattle or horses, were marked for "removal," or, plainly speaking, for murder. The expedition set out blithely enough, harboring no doubts of its complete success and not dreaming that any obstacle could be interposed to its formidable array. The first two "rustlers" encountered were found conveniently at their cabin doors and promptly despatched, though they died with their guns in their hands and were able to make a feeble response to the overwhelming numbers. But beyond these two assassinations the expedition was unsuccessful. The small settlers throughout the region were in sympathy with the men marked for death. The news of the "invasion" spread with incredible swiftness, and before the expedition could reach the homes of other intended victims the "rustlers" and their farmer allies, under the aggressive leadership of Jack Flagg—a noted character in the neighborhood—rallied in large numbers. They surrounded the "invaders" at a farm-house, and would have exterminated them to the last man except for the timely arrival of a troop of United States cavalry from the nearest fort. After several months of delay, the powerful political influence of those who had organized the expedition succeeded in setting its members free without serious punishment.

Public opinion differed much as to the justice of this bold effort to dispose once and for all of the annoying and costly evil of cattle-thieves. By some it was regarded as the irrepressible conflict between the irrigated farm and the free range. These thought that the real animus of the affair lay not in the just complaint against a few thieves, but in the fixed determination of those who profited from the unrestricted use of the public lands to prevent, at any cost, further settlement by honest farmers. On the other hand, there were many good citizens, men who had not hesitated to risk their fortunes in constructing irrigation works for the very purpose of opening certain valleys to settlement, who did not hesitate to

THE UNKNOWN LAND OF WYOMING

defend the expedition as the only possible means of ending an intolerable condition in the State. The writer has taken pains to gather testimony years after the event, when angry passions had wholly passed away, and found excellent evidence of the fact that those who were selected for extermination at the hands of the "invaders" were actually cattle-thieves; that it was clearly impossible either to end the evil or to stop its growth by appeal to the courts; and that farmers who settled in good faith were never molested by the large stock interests.

However, the political control of Wyoming speedily changed hands as the result of this dramatic episode. The party in power at the time of the event was voted into retirement, and the party which denounced the "invasion" as a savage and unmanly attempt to make widows and orphans of the wives and children of those who honestly sought homes in the public domain was installed in the Capitol at Cheyenne. The probable truth of the matter is that wealthy cattlemen had a real grievance which they could not adjust peacefully without years of patient waiting. They felt perfectly justified in their consciences in resorting to violence. They believed the result would be favorable to the prosperity and good name of the State. This actually proved to be the case, but in a very different way from what they had anticipated. It drew attention in a startling manner to certain evils inseparable from the open range and put these evils on the road to ultimate settlement through Congressional action. It broke the power of what was doubtless justly known as "The Cattle Ring" in State politics. It gave an impulse to better forms of develop-

THE CONQUEST OF ARID AMERICA

ment and a healthier tone to public thought. Above all, it taught the men of the frontier the great lesson that this is a government of laws and institutions, and that nothing is to be gained in the end by resorting to violence, at least when nothing more precious to humanity than the ownership of dumb brutes is the issue involved.

The irrigation development of Wyoming is distributed over a wide area. As has already been said, it has grown up mostly as an adjunct to the cattle business. The water supply is very abundant, and admitted of the construction of many cheap canals by settlers, without the assistance of outside capital. Grass, grain, and vegetables are the principal crops, but the State annually sends from half a million to one million dollars beyond its borders for agricultural products. This is due in part to the fact that the chief farming centres are widely separated from the principal towns and not connected with them by railroads. It is due also to the fact that small-farming has not yet been undertaken to any extent, and that farmers produce mostly only what they can feed to cattle or sell to others having cattle to feed.

The most active agricultural region is in the north-central portion of the State, in Johnson and Sheridan counties. It was from this district that the marvellous wheat, barley, and oats were sent to the World's Fair at Chicago—products which astonished Eastern farmers and won the highest prizes. Here, as indeed throughout the State, the farmers are highly prosperous. They have never known the miseries of their drought-stricken neighbors so close at hand in Nebraska and Dakota. Selling their product at home, they have not felt the bur-

THE UNKNOWN LAND OF WYOMING

den of transportation charges, nor had their prices much reduced by the glut of cereals in the world's market.

The earliest irrigation work of great importance was that at Wheatland, sixty-five miles north of Cheyenne. This was undertaken by local capitalists, headed by ex-Senator Carey. After surviving many difficulties, it at length entered upon a period of real prosperity and created the finest agricultural colony in the State. It is interesting to note that many of its people represent the overflow of the famous Greeley Colony in neighboring Colorado.

The most notable recent enterprise in Wyoming is that undertaken in the Bighorn Basin by the famous scout, William F. Cody, familiarly known as "Buffalo Bill." This energetic and ambitious man, who has twice won fame—first as a daring and successful scout, and then as exhibitor to two continents of the life, people, and customs of the Wild West—has laid broad and deep the foundations of a still stronger claim to remembrance. He conceived the idea of planting civilization in one of the wildest regions which he had first known as hunter and Indian-fighter. The money which the public poured into the coffers of his Wild West show, Cody used in reclaiming and colonizing a large tract in the valley of the Shoshone River in northern Wyoming, twenty to sixty miles from the Montana line and immediately east of Yellowstone Park. The altitude here is about four thousand feet, and the climate suited to the production of diversified crops, including hardy fruits. It is also the finest of cattle countries, and is surrounded by an abundance of mineral and timber. Its products find ready sale in the large and growing mining camps of the neighborhood, as well as of Montana. In time the region must acquire a large population, supporting a many-sided life, and form a very substantial monument to William F. Cody and his work for the West.

These lands were among the first to be investigated by the Reclamation Service, and now form the "Shoshone project" which is described elsewhere in this book. As there stated, Colonel Cody turned over to the Government all his rights in the river, in order that the project might be developed on the largest possible scale.

In the southern part of Bighorn County, a tract of thirty-five thousand acres of alluvial land is being reclaimed. The water is taken from the Bighorn River, which furnishes an abundant supply. A large steel flume to carry the canal across the river is a notable feature. The elevation of the land is about four thousand feet, and the climatic conditions favor the growth of fall wheat and barley, corn, and vegetables. The nearest town is Worland. The great drawback to the development of this region at present is the lack of railroad facilities. The settlers expect extensions of the railroads from Cody and Casper to Thermopolis, in 1906.

The largest project by private capital in Wyoming now is one for the reclamation of about two hundred thousand acres of land between Cody and Burlington, in Bighorn County. The water is to be taken from the south fork of the Shoshone River, in the Oregon Basin. These lands have been segregated under the "Carey Act." It is expected that the lands will be open for settlement in 1907.

All the public lands in Bighorn County are in the

THE UNKNOWN LAND OF WYOMING

Lander United States Land Office District, except a few townships in the eastern part of the county, which are in the Buffalo Land District.

Considerable irrigation development by private enterprise is also taking place in Fremont County. About twenty miles north of the town of New Fork, a tract of seven thousand acres on the western slope of the Wind River Mountains is being reclaimed and will soon be open to settlement. A canal has also been completed for the irrigation of a tract of over six thousand acres in the western part of this county, on a fork of the Green River, and the land is now open to settlement. When this county is provided with railroad facilities, a vigorous and healthful development may be expected.

The great activity of irrigation development in Wyoming is shown by the fact that for the two years ending November 30, 1904, the State Engineer's office issued 1109 permits to appropriate water, and 316 applications for enlargement of water appropriations were filed in the same period.

In the decade between 1890 and 1900, Wyoming's population increased from 60,705 to 92,531—over fifty per cent.—and the present population is estimated by the Governor at over one hundred and twenty thousand.

In the sheep industry, Wyoming ranks first in the United States, having now over five million sheep. The wool clip of 1900 amounted to 27,119,718 pounds, with an estimated value of $3,796,760. The value of its cattle in the same year was over six million dollars. Horses raised on its elevated table lands are superior in wind and endurance. In May, 1903, the saddle-horse "Wyoming," considered a typical product of the State, was presented to President Roosevelt, and is now in the White House stables.

In 1904, Wyoming's production of coal amounted to over five million tons and gave employment to over nine thousand men. One-fourth of the area of the State is underlaid with coal. It is estimated that since 1883, the State has produced 15,206,092 pounds of copper, of the value of $2,267,775.60. The mining of iron ore is assuming considerable importance, as well as the production of petroleum. The State now has nearly five hundred manufacturing establishments, representing an investment of over three million dollars. There are forty million acres of government land and about ten million acres of timber. With such prodigal wealth of raw materials, such solid and substantial values already developed, and such an impulse toward growth, Wyoming is destined to be the home of many millions of prosperous and contented people.

The irrigation law of Wyoming is widely celebrated and has been influential in moulding the institutions of other States, and even those of Canada and Australia. It is based on the sound proposition that water belongs to the public and that only the public can grant the right to its use, which must be a beneficial use, with due regard to the rights and interests of all other users, present or prospective. This fundamental idea is applied by means of a thorough and effective administrative system. No canal proprietor is permitted to open or close the headgates of his own canal. This is done by public officials clothed with police powers, who divide the water in

THE UNKNOWN LAND OF WYOMING

strict conformity with the decrees of a State Board of Control. The result is an almost complete absence of litigation and of unseemly strife among neighbors which mar the irrigation industry in many other localities.

But while the Wyoming method is admirably adapted to that State, where a multitude of individual proprietors take water from a common source, it does not meet the needs which are arising with new conditions throughout the West. There is a strong and irresistible tendency toward the consolidation of many small conflicting works into comprehensive public systems, and there can be no doubt that on each watershed there will finally come to be one system, diverting all the water that may be diverted, storing all that may be stored, and pumping all that may be pumped. This method will make for the highest efficiency and greatest economy. Each watershed will be a unit in itself and its affairs will be administered by its own people—a democracy of landed proprietors owning both the water and the soil—without State interference. The Wyoming system is a vast improvement over the chaos which prevails where each man is a law unto himself, but the new system which is arising under the influence of national irrigation represents a step **far in advance of** the Wyoming system.

CHAPTER IX

THE PROSPERITY OF MONTANA

MONTANA is a State of magnificent resources. The first white men who ever saw it — French explorers in the middle of the eighteenth century — called it "The Land of the Shining Mountains." The appellation is true as well as poetic, for it is the possession of its snow-capped ranges, reflecting the light of the brilliant sky, which differentiates Montana from the adjoining prairie States of the Northwest. It is the mountains which hold the wealth of waters and minerals and make the character of the climate.

Montana ranks third in point of area among American States, and third in the value of its annual mineral output. It is yet too early, by many years, to estimate its final place in extent of population and agriculture. To-day mining is the first of its industries, stock-raising the second, agriculture the third. Mining gave the impulse to its settlement and is the backbone of its prosperity. The forty millions of dollars annually taken out in copper, lead, gold, and silver make it one of the most prosperous of western communities. The discovery of new mining districts steadily continues, and the flow of wealth from this item of the State's resources will endure indefinitely. The conditions of the stock industry are

THE PROSPERITY OF MONTANA

very similar to those which we observed in Wyoming. Of the total population—about three hundred thousand—the farmers are a small minority. Nevertheless, irrigation is recognized as one of the most important interests of the State, and the field open to settlement offers many attractions.

The first ditches in Montana were made for the purpose of washing gold-bearing gravel along the bars and gulches. When their usefulness in this direction was exhausted they were turned into irrigation canals by the farmers who came close upon the heels of the early miners. For many years development was limited to works of this humble character. Farmers had their own individual ditches, or combined their labor in making canals sufficient to water small districts. In this manner most of the mountain streams capable of easy diversion were utilized. As in Wyoming, irrigation was largely used as only an adjunct to stock-raising. In recent years legitimate agriculture has begun to make rapid progress. Large capital has been invested in a few comprehensive irrigation systems, notably in the valleys of the Dearborn and the Sun rivers, north of Helena.

Montana is divided into three natural drainage areas—those of the Missouri and Yellowstone rivers on the east of the main range of the Rockies, and that of the waters tributary to the Columbia on the western slope of the mountains. The eastern slope embraces the fertile valleys of the Yellowstone, the Gallatin, the Madison, the Jefferson, the Beaverhead, the Prickly, and the long valley of the Missouri, with the Milk-river system in the extreme north, on the border of Canada. The western slope is mountainous and heavily timbered, with comparatively small though fertile valleys. The principal streams are the Flathead, Clark's Fork of the Columbia, and the Kootenai. The ultimate extent of irrigable land within the boundaries of Montana is purely speculative, estimates ranging from ten to thirty million acres. In the matter of water supply the State is among the most fortunate in the West, though its full utilization will require vast expenditure for the construction of storage-works and of long canals. Some of the largest rivers, like the Missouri and the Yellowstone, are enclosed by high bluffs, and water can be taken to the elevated plains, comprising the larger areas of valuable land, only by means of diversions made high up upon the streams.

The opportunities which Montana offers to settlers have not been appreciated as they deserve. This is doubtless due to the severity of the climate, which is generally misunderstood. The State is in a high latitude, and does, indeed, experience cold winters. But its valleys are comparatively low, averaging much lower than those of Wyoming, Colorado, Nevada, and Utah, and its climate decidedly healthful. The thermometer goes twenty or thirty degrees below zero in the winter, but this degree of cold in the dry air of Montana is much less disagreeable than ten degrees above zero in any of the cities on the borders of the Great Lakes. On the other hand, the State enjoys a remarkably even prosperity, and no other localities offer better certainty of home markets, where the products of the farm can be disposed of at good prices.

There are many large and growing towns, and two or three cities of considerable size. The mining popula-

THE PROSPERITY OF MONTANA

tion is destined steadily to increase, while manufacturing must begin in earnest during the next decade. The wheat, rye, oats, and vegetables produced in the irrigated valleys are remarkable both in quantity and quality. The brewers of Brooklyn, New York, selected the Manhattan Valley for important agricultural operations, because they found it would grow the finest barley in the world. Small fruits are prolific and of fine flavor.

Even the orchard fruits, especially apples and plums, are produced successfully in the more sheltered valleys. The exhibits which one sees at county fairs, particularly at those on the western slope in valleys like the Bitter Root, make surprising revelations of the fruit possibilities in this northwestern State. But the settler's chief opportunity will be found in supplying the common farm products required by the large and growing population in the cities and towns. Of the present local consumption, the great portion of the pork, bacon, ham, lard, and cheese, and nearly half of the flour, butter and eggs, are now brought in from elsewhere. Efforts are being made to change these conditions, and especially to increase the area cultivated in hard wheat. When such facts are considered in connection with the cheap land, abundant water supply, and healthful climate, it is apparent that Montana offers great attractions to colonists.

The Gallatin Valley, southeast of Helena on the main line of the Northern Pacific, is the most famous agricultural district of Montana. It is well settled, with a class of thrifty farmers engaged in producing a variety of ordinary crops. Bozeman, the county seat, is the home of the State Agricultural College, and this institute has done much to raise the standard of irrigation and of farming in the locality, and thus to enhance the valley's prestige. The Missouri Valley, in the neighborhood of Great Falls, and the Bitter Root Valley about Missoula, are other well-developed districts. Crops are generally planted in April or the first half of May, though sometimes in March. The spring rains continuing until the middle of June, irrigation does not begin until that date. Cattle, sheep, and wool are shipped to eastern markets, but other products are consumed within the State.

While copper and the precious metals are the chief mineral products of the State, it is rich in lead, iron, coal, building materials, and precious stones. It is estimated that an area of not less than fifty thousand square miles is underlaid with bituminous or lignite coal of good quality. Coke is a growing product. The State is also rich in forests and abundantly supplied with natural water-power. It has, in a word, all the materials of a diversified industrial life.

The social and political life of Montana is vigorous and interesting. Both the climate and the industries are calculated to breed a sturdy and self-reliant people. Helena, the capital, located in what was formerly known as Last Chance Gulch, has long enjoyed the reputation of being the richest city in proportion to its population in the world. Butte is still larger—the largest mining camp in the United States. These two leading towns present radically different aspects of western life. Helena is the political and social capital, Butte the grimy centre of industry. Both have enjoyed phenomenal prosperity, and escaped, to a large degree, the relapses which have afflicted other ambitious western cities at various times.

THE PROSPERITY OF MONTANA

The area where crops can be successfully grown without irrigation is small, and practically none is now left open to entry. Good land with water sells for from ten to thirty-five dollars per acre, the average price being about twenty-five dollars. Improved lands bring as high as seventy-five dollars per acre.

The truth is that Montana has been, and is yet, a marvellously substantial State. It has enjoyed a steady stream of wealth from the mine, the range, and the farm. Its mercantile enterprises have naturally thriven under these conditions, and labor has been busy and well paid. It has not been the policy of the people to encourage immigration on reckless lines merely to increase the population. On the contrary, the public sentiment has been notably conservative, and has only urged those to come who could be self-supporting by tilling the soil or establishing other industries.

Great Falls, located at the most eligible water-power of the Upper Missouri river, has enjoyed a remarkable growth of population, and promises to become in time one of the great cities of the West. In addition to the water-power, it has the advantage of being surrounded by the raw materials of manufacture, in the shape of coal, iron, timber, and the products of the range—such as wool and hides—while large agricultural districts are tributary to it. There are many important towns along the line of the Northern Pacific and the Great Northern railroads. Of these Missoula is a prosperous mercantile point on the western slope, and Billings is the focus of agriculture in the Yellowstone Valley.

Viewed as a whole, Montana is a State of substantial achievement and of splendid promise.

CHAPTER X

THE AWAKENING OF NEW MEXICO

IN the southwestern Territories modern methods of reclamation are asserting their influence in the midst of historic and prehistoric irrigation scenes.

In 1539 Fray Marcos de Nija, the earliest European who trod the soil of New Mexico, travelled for five days through a "valley well watered and in a high state of cultivation, so that three thousand horsemen might have been sustained there." Another sixteenth-century visitor saw corn-fields "watered by a small river which flowed near by, along the banks of which were growing great beds of roses, similar to those of Castile." Many a tourist on the Atlantic and Pacific Railroad has seen the industrious Pueblo Indians at work in their fields about Laguna. The travellers of three hundred and fifty years ago looked upon these same fields, which were irrigated then precisely as now, and as they probably had been for centuries before.

New Mexico is much less favored in its water supply than the northern States of the arid region. Many of its streams are torrential and intermittent in character, carrying water in floods at some seasons and exhibiting dry channels when moisture is most needed. A large portion of the water supply, when the irrigation indus-

THE AWAKENING OF NEW MEXICO

try shall be fully developed, will be obtained by storage and from underground sources. This process has already begun, but its operations will be much extended. Scattered all over the territory are the petty ditches of that numerous Mexican and Indian population which lives in serene peace and comfort upon the fruits of its unambitious efforts at tilling the soil.

The important streams are the San Juan in the northwestern corner of the Territory, the Rio Grande, which flows through the central portion from Colorado to Mexico, and the Rio Pecos in the southeast. These streams and their tributaries furnish the basis of the modern irrigation industry of New Mexico.

The northwestern part of the Territory is a picturesque and promising region, fortunate alike in mineral and water resources, in the fertility of its soil, and the charm of its climate. A number of small irrigation systems have been constructed, but storage will be required before the opportunities of the district can be extensively realized. The rivers are the San Juan and its tributaries, the most important of these being the Pine, the Animas, and the La Plata. When these are fully utilized, thousands of small farmers will be able to establish profitable industries, including the culture of finely flavored, delicate fruits. They will find home markets in surrounding mining camps and in supplying feed for sheep and cattle which range upon the public pastures. Although this portion of the Territory is now remote from the main lines of railroad travel, its superior advantages must attract the attention of enterprise and immigrants in the future and make it one of the most prosperous parts of the future State.

THE CONQUEST OF ARID AMERICA

New Mexico is distinguished by large land grants dating from the days of Spanish control. They were naturally located along the watercourses, in what appeared to be the most attractive portions of the field open for selection. These enormous grants have materially retarded development, for the reason that their titles were in dispute for many years and their owners generally "land poor." The Court of Private Claims, which recently completed its work, performed a great service for the arid region in the settlement of such disputed land titles.

One of the most important of these properties is now known as the Maxwell Land Grant, and constitutes a principality in the northwestern part of the Territory, encroaching slightly upon Colorado. Large capital has been used in the development of the mineral and agricultural resources of this grant. Its principal streams are the Vermejo and the Cimarron. Both have been utilized extensively in connection with systems of reservoirs and canals which are notable for some of their engineering features. Large areas have been irrigated and are cultivated in various crops.

The waters of the Rio Grande have been diverted at many points along its course. This river rises in Colorado, where a large portion of its waters are taken out for use in the San Luis Valley. This interferes with New Mexico irrigation during the stage of low water in the summer. The stream reaches old Mexico still further diminished, and vexatious interstate and international complications have long existed from this cause.

In 1896, the Republic of Mexico made formal complaint of the diminished flow of the stream at El Paso.

THE AWAKENING OF NEW MEXICO

Its citizens have made a beautiful agricultural and horticultural region where Texas reaches out a slender finger of prosperity at El Paso. Here, they have practised irrigation for over two hundred and eighty years, and live with an enviable degree of comfort and thrift, though their methods are crude and ancient. The matter was referred to an International Boundary Commission, which recommended the construction of an international dam in the stream, at a spot called Elephant Butte, about four miles above El Paso, the extension of Mexico territory up to the proposed dam, the equal ownership of the reservoir and water supply between the two republics, and the construction of other large reservoirs on the stream in New Mexico to be forbidden.

There were many objections to this plan. New Mexico objected because some twenty-five thousand acres of good land belonging to its citizens above the dam site would be submerged, because of the prohibition of other dams on the stream, and for other reasons. Texas also objected. At the Twelfth National Irrigation Congress, held at El Paso, Texas, November 15 to 18, 1904, announcement was made of a new plan for the solution of the difficulty by the engineers of the United States Reclamation Service, which immediately enlisted the enthusiastic support of the representatives of all the parties in interest. A resolution endorsing this new plan was signed by the delegates from Mexico, New Mexico, and Texas, and the way now appears to be clear for the settlement of the question in a way to do justice to all.

The Reclamation Service has been surveying and studying the stream for two years with a view to developing its resources and settling this and other difficulties. Their plan is to build a storage dam at a new site about a quarter of a mile below Elephant Butte. It has been found that this can be done in such a way as to avoid the submergence of New Mexico's lands. The proposed dam will be 175 feet high at the lower end, the storage basin forty miles long, and the storage capacity two million acre feet, sufficient to irrigate one hundred and eighty thousand acres of land. It will be the largest dam on the river, and entirely interrupt the flow of the stream, even in wet years, and hold them for use as needed. It is such achievements as this which are demonstrating the wisdom of the policy of national control of the reclamation of the arid West.

Much of the most notable irrigation development in New Mexico is that which has been accomplished since 1890 in the Pecos Valley. It is in the southeastern quarter of the Territory, bordering upon the Staked Plains of Western Texas. No other locality in the arid region has had the benefit of such daring enterprise and dauntless faith as have been lavished upon this, originally one of the most forbidding and unpromising of Western valleys. By sheer force of money it has been translated from a semi-barbarous stock-range, fit only to support lean cattle, to an attractive field for settlement, where thousands of families can make their homes and win a certain livelihood from the soil. Civilization has laid its hand on the Pecos Valley and a crop of new institutions has begun to sprout from its soil.

The valley is fortunate beyond any other part of the Territory in its water supplies. The Pecos River and its tributaries drain a vast watershed and furnish a peren-

EAGLE DAM SITE, RIO GRANDE PROJECT, NEW MEXICO.

THE AWAKENING OF NEW MEXICO

nial flow of large dimensions. This has been conserved by large reservoirs, one of which is the second largest reservoir in the world. The valley is also blessed with extraordinary springs of flowing water, with artesian basins, and with underground supplies that may be lifted to the surface at comparatively small expense. With splendid disregard for immediate financial returns, these supplies have been utilized and led over the valley by a thousand miles of canals and ditches. The same liberal enterprise built railroads, established towns with modern facilities, and acquired large tracts of irrigable land. At Artesia, formerly Stegman, land is being rapidly taken up by settlers under the homestead and desert land laws, to be irrigated from artesian wells. The valley has passed safely through recent seasons of drought which proved disastrous to less favored sections, and land values are steadily rising.

Lying in an altitude varying from three thousand to three thousand five hundred feet, but in the latitude of the extreme south, the Pecos Valley enjoys a good climate. Its winters are short and not severe, though the mercury falls below freezing and thin ice is formed on still water. The summer days are hot, as elsewhere throughout the Southwest, but the nights are invariably comfortable, owing to the elevation of the country, which is on the high plateau of the Rocky Mountain region. The drawback about the climate is the wind, which sometimes develops into sandstorms of considerable severity. With the extension of the cultivated area and the growth of trees this disadvantage will be minimized.

Fields can be cultivated almost continuously and early

crops of vegetables and small fruits grown. Alfalfa is cut four times a year, and after that furnishes considerable pasturage. All grains do well, and Kaffir corn and Milo maize are extensively cultivated. Vegetables can be successfully grown throughout the valley, but seem to have been somewhat neglected in the south. Cotton raising is a flourishing industry here, and there is a gin at Carlsbad.

The Pecos Valley is particularly adapted to sugar-beet culture. A series of experiments has demonstrated that the soil and climate are especially favorable to the growth of beets. A sugar-beet factory was erected in 1896, and the farmers planted considerable areas to beets. The general average of all the beets delivered to the factory in car-load lots the first year, showed seventeen per cent. of sugar in the beets, with an average purity of eighty-two per cent. This is a higher percentage of actual extraction of pounds of sugar to pounds of beets than has been realized anywhere else in the world. Unfortunately, the factory failed and afterward burned down, so that the industry is now dormant. The people of the valley are full of faith in the industry, under proper management, and believe, apparently with good reason, that it is destined to become a " sugar belt."

The valley has not been in cultivation long enough to determine the limitation of its products. The chemical qualities of the soil have been the subject of careful study by experts, and the people are gradually learning to what uses different districts are best adapted. In the upper portion of the valley, in what is locally known as the Roswell country, there are several ranches which

THE AWAKENING OF NEW MEXICO

have been cultivated for many years. These have demonstrated beyond question the capabilities of soil and climate for the production of the finest apples, perfect in form, flavor and coloring. The raising of celery and cantaloupes are new and promising industries here. The lower valley seems more favorable to delicate fruits, such as peaches, grapes and apricots.

One feature of this valley is especially worthy of the attention of settlers. This is the fact that the best of pasture adjoins the irrigable lands, on either hand, so that fine cattle, sheep, and horses are profitably raised in connection with the small-farming industry. Raising winter fodder on their irrigated acres, the settlers co-operate in the management of their herds during the range season. For every acre under cultivation, there are three hundred acres of grazing land, over large areas of which an abundance of water is found at an average depth of from twenty to four hundred feet. Windmills dot the country, and practically every acre of gazing land is occupied by live stock.

The chief town of the lower valley, formerly Eddy, but now called Carlsbad, enjoys a considerable degree of prosperity, both as the centre of a flourishing agricultural region and as one of the many attractive health resorts. Roswell is the metropolis of the upper valley, and the scene of the greatest present activity of the Reclamation Service in the Territory. An account of the Government project on the Hondo River, near this place, will be found in the chapter on government reclamation work.

The development of the irrigation resources of New

THE CONQUEST OF ARID AMERICA

Mexico has been stimulated by a law enacted four years prior to the passage of the present national irrigation law, under which half a million acres of land were granted to the Territory on condition that they be reclaimed by private enterprise. These lands are selected by a Land Commission, which in 1904 had selected over 233,000 acres. The disposal of these lands is in the hands of a Commission of Irrigation, created by the Territorial Legislature in 1901 for the purpose of carrying out the provisions of the act. By the report of this Commission for 1904, it appears that there were then pending before it six private irrigation enterprises with a total estimated capacity of 157,000 acres, and two others not yet acted upon, with a total capacity of 42,000 acres. No work had been done, but it was believed that within a year several of these irrigation projects would be undertaken and work on the necessary dams, reservoirs, and ditch systems be commenced.

The resources of New Mexico, while probably not as rich as those of more northerly states, are yet diversified and largely undeveloped. The annual output of gold and silver is increasing, and seems likely to continue indefinitely to do so. A fine quality of coal is found in large quantities, and is an important item of regular income. The area of merchantable timber is said to amount to five million acres, and that of woodlands, useful for fuel and fencing, is much more extensive. There are four forest reserves with a total area of 5,125,000 acres. The mining of precious stones, which dates back to the Spanish conquest, is a flourishing and growing industry. The turquoise mines are particularly rich and profitable.

THE AWAKENING OF NEW MEXICO

Though the amount of their production is closely guarded, it is known to be large, while the quality of the stone is equal to that of Russia, Persia, and the East Indies.

The social fabric of the Territory is a curious blending of Mexican peons, of town-building Indians, of hardy frontiersmen engaged in mining and stock-raising, and of enterprising newcomers who believe in the future of the country. Of these elements, the Mexicans are the most numerous. They do not differ materially from their kinsmen on the southern side of the Rio Grande. Living in scattered settlements along the mountain streams, they enjoy a comfortable existence in return for their humble labor. The Indian population includes the Pueblos, the Zunis, the Navajos, and the Apaches, and is marvellously interesting, and usually peaceful and industrious. The condition of these Indians is being slowly improved by the construction of irrigation works for their benefit, and other means. The growth of the white population has been slow, but will increase rapidly with the development of irrigation.

New Mexico is one of the American communities whose greatness is of the future. Well endowed with raw materials, it awaits the impulse to be imparted by the new century and the pressure of an outreaching civilization. It is distinctly a land of opportunity, and one of the few remaining spots where settlers of small means can hope to prosper now, without awaiting the slow development of resources by the National Government. A Bureau of Immigration is maintained by the Territory, for the purpose of supplying information to intending settlers. The Secretary's address is Santa Fe.

CHAPTER XI

THE BUDDING CIVILIZATION OF ARIZONA

ARIZONA is a land apart. With the single exception of southeastern California, it differs in many respects from all other sections of western America. This is especially true of all those portions of the Territory which will sustain the densest future population and develop the characteristic institutions of the country.

Speaking of its atmosphere—the product of its peculiar climatic conditions and physical environment—Whitelaw Reid has said: "It seems to have about the same bracing and exhilarating qualities as the air of the Great Sahara Desert in northern Africa, or of the desert about Mount Sinai, in Arabia. It is much drier than in the part of Morocco, Algiers, or Tunis usually visited, and drier than any part of the valley of the Nile north of the First Cataract. It seems to me about the same in quality as the air on the Nile between Assouan and Wady-Halfa, but somewhat cooler."

This description of the Arizona air, which is remarkably happy, may be accepted as a key to the true character of the country. It is a semi-tropical desert, like the region about the southern and eastern shores of the Mediterranean, where civilization was born of the ancient art of irrigation. This is said with reference to the southern and western parts of the Territory, which are drained by the Gila and Colorado rivers. Northern Arizona is distinguished by its mines, its notable forests, and the indescribable grandeurs of the famous Colorado canyon. The southeastern quarter, which adjoins New Mexico, is a great pasture, bearing scanty or generous crops of nutritious wild grasses, according as the season is dry or wet.

The Salt River Valley is the glory of Arizona. Approaching it from either of the transcontinental railways the traveller sees naught but the gray desert soil, marked by the gnarled branches of the mesquite and the slender pillar of the cactus. Even the mountain-sides appear to be devoid of verdure and tanned to a dark brown by the sunshine of centuries. But suddenly all the beauties of the Garden of Eden burst upon the astonished gaze of the visitor. Wherever the waters of irrigation have moistened the desert, and man has planted the seed of grass, flower, or tree, the most luxuriant vegetation has sprung from the soil to revolutionize the appearance of the country.

The capital city of Phœnix—risen from the ashes of a forgotten people—is the pulsating heart of the new life of Arizona. Here are modern business blocks, handsome public buildings, busy stores, a promising university, and hundreds of beautiful homes resting under the shade of palm, magnolia, and pepper-trees. Tucson and Yuma, though thriving and populous, are Mexican in architecture and habits. Prescott, Flagstaff, and numerous other communities in the higher altitude are the products of the mining industry. But Phœnix is distinctly modern, and almost wholly the offspring of irrigation.

THE CONQUEST OF ARID AMERICA

The Salt river is the largest tributary of the Gila. It has been the scene of active irrigation enterprise since 1867, but particularly during the last ten years. It is an interesting fact that the works first built followed the lines of prehistoric canals. Reclamation has been extended to both sides of the valley, but cultivation is oldest and much the most extensive on the northern side, around Phœnix. Here a number of canals were consolidated into a single system, the managers of which have made improvements and extensions year by year, and gradually evolved a work of great perfection and completeness.

On the south side of the river a similar consolidation has occurred. Here settlement was begun in 1878 by Mormon colonists, who founded the charming place now known as Mesa City. There are several independent irrigation systems upon this side of the valley, the most important of which is the Highland Canal, which runs along a high level and waters thirty thousand acres of valuable land. Water-power is obtained in connection with the irrigation canals on both sides of the valley, and electrical power is applied both to lighting and transportation.

Tributaries of the Salt river flowing from the mountains on the north, notably the Rio Verde and the Agua Fria, will furnish water for new and large enterprises. Storage is the feature of these works, and reservoirs have been constructed in a number of instances. Both on the upper and lower courses of the Gila river important irrigation canals are planned, and a number have been completed. Much difficulty has been experienced in building enduring dams along this erratic

THE REDEEMED DESERT IN ARIZONA.—1. Harvesting Third Crop of Alfalfa, near Yuma. 2. Irrigated Barley Crop, Yuma Valley.

stream. Sudden and powerful floods sweep down the valley during the season of melting snows, and it is the nicest engineering problem to make constructions which will stand the test.

In the first edition of this book, issued in 1900, I said of the valleys of the Gila and Salt Rivers: "The agricultural districts suffer for lack of water during the dry summer season, when water is most needed. The only possible solution of the problem will be construction of large reservoir systems at the mountain sources of the streams. Nature has provided phenomenal facilities for such storage works, but the opportunity has not been utilized, owing to the large cost involved and to the fact that no single company could afford to make improvements which would be equally beneficial to all who draw supplies from these streams. The work is of such importance as to justify an expenditure of public money, especially as large areas of public lands would be made habitable in consequence."

The truth of these statements has long been understood by well-informed men in the Territory. Thus Governor Brodie, in his Annual Report to the Secretary of the Interior for 1904, says: "To-day we are forced to admit that there is not a valley in the Territory where land is available to the homesteader where he can settle down to the successful pursuit of agriculture with the independence of a Mississippi Valley farmer. . . . With a great system of water storage established in Arizona all public land susceptible of irrigation will be at the disposal of the husbandman. There will be land for millions where there is to-day scarcely enough for the present population. Arizona contains more than ten million acres of land that can be cultivated."

This vital need of water storage is now being supplied by national enterprise in the Salt River Valley and on the Colorado River. The interest of the people in irrigation development was shown by the enactment of a law authorizing the bonding of counties for the construction of public storage works, and by the creation of a Territorial Water-Storage Commission, prior to the passage of the national irrigation law. The average annual precipitation in the Salt River Valley for the past twenty years was only 6.9 inches, which makes farming without irrigation impossible. The passage of the national irrigation law and the prompt and intelligent action of the officials of the Reclamation Service greatly pleased the people of Arizona, who see in this development under the competent guidance of the Nation a lively hope of becoming a great agricultural community. The lands capable of being reclaimed are not alone the many fertile valleys, but the elevated table-lands as well, which are very rich and of enormous extent. A sum far in excess of the total amount now in the reclamation fund could be expended in this Territory, with the result of making homes for thousands of prosperous people.

The climate of Arizona varies widely with different altitudes. In those portions of the Territory most favorable to settlement, including the Salt River and Gila Valleys, newcomers find the summer heat somewhat trying. Old settlers, who know how to adapt themselves to their environment, do not find the heat oppressive; but it is something for a newcomer to take seriously into account. In the more northerly portions of the Territory, however, the climate is wholly different and the disadvantage of extreme heat far less. Throughout the whole of the Territory, the winters are delightful.

In the Salt River Valley, all classes of fruit have been tested sufficiently to furnish reliable conclusions as to the range of production. The climate is semi-tropical and the products similar to those of the lowland districts of California and the region around the Mediterranean.

Government reports show that the highest and lowest temperatures at Phœnix averaged for eight years, as follows: November, 78½ and 42; December, 73½ and 36½; January, 65½ and 32; February, 71½ and 35½; March, 81½ and 41; April, 86½ and 46. Orange trees successfully withstand a temperature of 28° above zero. Hence, it is no surprise to find them growing successfully in the Salt River Valley, at Yuma, and elsewhere in central and southern Arizona. The determination of the exact limits of the citrus belt is a nice problem in any country. A certain elevation above the river, and a certain amount of protection from the wind and from the rising sun, are essential. The most favored spots are usually those which are screened from the first rays of the morning sun by a background of eastern hills. This condition permits a gradual warming of the atmosphere, so that if there has been a slight frost during the night no serious harm is done to fruit or tree.

Wherever oranges can be grown at all, the area suitable for their production is likely to be exaggerated by those who sell climate by the acre. While the orange districts of Arizona are not as yet perfectly defined, there is no longer any question of the production of citrus fruit nor as to its quality and the early date at which it ripens. It anticipates the Southern California crop in the market, though not the crop of Northern California, which is several weeks ahead of the southern product.

Wherever the orange can be cultivated, the less tender semi-tropical fruits—figs, olives, almonds, pomegranates—may be certainly counted upon to grow even more surely and over a large area. The largest fig orchard in the United States, and one of the largest in the world, is located in the Salt River Valley. This industry has not yet proven profitable, either in Arizona or California, speaking broadly, for the reason that our people have not all mastered the art of curing and packing, though much progress has been made at Fresno. The other products mentioned are thoroughly successful. So also are the finest qualities of raisin, wine, and table grapes, and of the deciduous fruits, such as peaches, apricots, prunes, pears, and apples. Yuma lays down table grapes in San Francisco before the California product is in the market. With better railroad facilities and rates, Arizona would be a strong competitor of Florida and the West Indies in the shipment of early vegetables to eastern and northern markets. At the government experiment station near Tempe is a date orchard of eleven acres, and there are strong hopes that this new industry will take root and prosper throughout the southern portion of the arid region.

The major portion of the irrigated land is tilled in large farms devoted to grasses and cereals. Alfalfa is the favorite fodder crop, and the valleys are great feed-

IRRIGATING A YOUNG ORCHARD BY FURROW METHOD, ARIZONA.

BUDDING CIVILIZATION OF ARIZONA

ing grounds for cattle, horses, hogs, and sheep. While stock-raising is important, it is in a less prosperous condition than formerly, on account of prolonged droughts, especially that of the year 1903, which caused great loss. As irrigation development progresses, the raising of live stock on large areas will give way to the intensive cultivation of small areas, to which the conditions of soil and climate are extremely favorable.

Ten acres in southern Arizona constitutes a good-sized farm. Variously planted to vegetables, small fruits, orchards, and grass, and cultivated by the most modern methods, such a farm should yield a far better living and make a surer provision for old age than one hundred acres in the Eastern or Middle States, which depend upon rainfall, and consequently produce the cheaper class of crops.

Arizona is very rich in minerals and its development has brought a considerable degree of prosperity in the past few years. Copper is now the principal product, having made the phenomenal growth of from two million pounds in 1880 to two hundred and thirty million pounds in 1904. In 1903, the Territory held third place in copper production, being exceeded only by Michigan and Montana; it now claims the second place, and, if the great Cananea mines, situated a short distance over the boundary line in Mexico and virtually dependent upon Arizona for their development, were included, it would be entitled to first place. The value of the product for 1904 was nearly thirty million dollars, and the larger portion of it was manufactured in the Territory. The production of gold, silver, and lead is also large, and rapidly increasing.

THE CONQUEST OF ARID AMERICA

Although Arizona is popularly regarded as a treeless region, it counts among its natural resources one of the largest forests in the world. This is the "Mogollon Forest," covering an area of ten thousand square miles. There are eight forest reserves in the Territory.

An interesting industrial development of late years is the raising of ostriches. In the Salt River Valley, near Phœnix, it has been demonstrated that they will thrive and produce feathers profitably. There are now more than sixteen hundred birds on this farm, which feed contentedly in the alfalfa pastures.

Lacking nothing in general advantages, Arizona has suffered from the popular belief that there is a deficiency of the higher forms of industrial and social development which have made portions of California the paradise of the common people and which are now rapidly shaping the institutions of the arid region. That this impression is passing away is evidenced by the fact that Arizona's population increased 104.9 per cent. between 1890 and 1900, and is still growing rapidly. The broad foundations of an intense economic life have been well and substantially laid, and the superstructure is rapidly rising. As the work progresses more and more, it will be understood how little the country has been appreciated, and it would be rash to attempt to predict its future greatness.

The people of Arizona have been drawn from many different sources and from more than one race, but the pushing American element is distinctly dominant. While there are many lower-class Mexicans, they are much less numerous here than in New Mexico, and less widely diffused over the Territory. The Indians, who are seen everywhere, even in the best settled districts, are inoffensive and usually industrious. Like the Mexican peons, they are useful laborers in the simpler agricultural and manufacturing tasks. There are many tribes, some of which were warlike in recent years, but these are now kept closely confined to their reservations and no longer constitute a menace to settlement.

Arizona has developed a spirit of intense local pride which bodes well for its future greatness. It is a good recommendation for any country when those who know it best exhibit the most confidence in its future. It is to be hoped that the energetic and optimistic people of Arizona may realize their high ambitions, including their laudable desire for statehood.

The settler intending to go to Arizona will find his best opportunities under the progressive development of the United States Reclamation Service. The opportunities for private enterprise, except to large capital, are practically exhausted.

Part Fourth

THE TRIUMPH OF THE MOVEMENT

"The passage of the National Irrigation Law is one of the great steps not only in the progress of the United States, but of all mankind. It is the beginning of an achievement so great that we hesitate to predict the outcome."—*Theodore Roosevelt.*

"It is men with hearts who have done it; men with imagination, illumination, prophecy, conscience. The fact that it pays is important, but it is secondary. If the business argument could not have been sustained, the movement would have died, but without the moral force the business argument would have shriveled like a leaf in the sand. The architects and builders of this great plan of redemption are and have been men of heart as well as brain, men of tact and of love for humanity, as well as men of firm convictions and shrewd business sense, men who look on an acre of land or a gold coin merely as a token to be used for the betterment of humanity."—*El Paso, Texas, Herald, June 17, 1905.*

JOHN WESLEY POWELL.—First Scientific Explorer of the Arid Region.

CHAPTER I

THE RISE OF A NEW CAUSE

THE true history of irrigation in America would involve a comprehensive study of the life of the Western people during the past two generations, with a study of certain communities which trace their civilization to a period much more remote. The facts for such a history would be found in the record of exploration and colonization, in the expansion of pioneer camps into villages, cities, and States, in the evolution of a multiplicity of laws and judicial decisions concerning land and water. For irrigation is the life-blood of institutions in the Western half of the continent and its history is, in a marked degree, the history of the people themselves. In this chapter the subject will be sketched in relation to one aspect only,—the birth and progress of the organized movement which finally led to the adoption of a new policy of internal improvement by the United States.

First on the roll of irrigation champions stands the name of John Wesley Powell. He was a soldier, a poet, a scientist, a lover of his kind, but in no sense a man of practical commercial instincts. It is worth while to note this fact, because the West owes much to men of another type—men who saw the opportunity to make great fortunes in the development of the country and whose business enterprises were conceived and executed upon so magnificent a scale that they are entitled to remembrance among the builders of empire. Collis P. Huntington, James J. Hill, William J. Palmer, William A. Clark, and a hundred others, wielded large capital, built railroads, opened mines, and prepared the way for millions. Brigham Young organized a host of settlers, turned the mountain streams upon the soil of the desert, directed the growth of towns, industries, and farming communities. These, and ten thousand men of lesser achievement, were of the practical sort indispensable to the work of transforming a wilderness into a seat of civilization.

Major Powell was entirely different and, perhaps, the most distinguished type of another class which has contributed in its own way to the conquest of Arid America. He did not invest money, but he invested ideas. He was not interested in making a fortune for himself, but in making a fortune for the Nation, for humanity. He saw the West with the eyes of a prophet and, with splendid imagination, beheld not only the opportunity which awaited a great people, but the measures which must be adopted to take best advantage of the opportunity. He was the kind of man who is commonly regarded as an enthusiast and a visionary until his dreams come true—the kind of man, as Elbert Hubbard said of another, doomed "to become rotting logs which will nourish banks of violets." It should not be inferred that he accomplished nothing in his lifetime. On the contrary, he accomplished much, but it was in the nature of preliminary work. The great results he was not given to see with mortal eyes.

THE RISE OF A NEW CAUSE

After the Civil War, in which he had borne a conspicuous part, Major Powell became geographer, geologist, ethnologist, and explorer of the arid region. His daring descent of the Colorado River is one of the historical episodes of the Far Southwest, but it was no deed of idle heroism. He was engaged in looking the country in the face, that he might know what it held for the future of men. His report entitled, "The Lands of the Arid Region," was one of those rare public documents which becomes classic literature. His preliminary work in the examination of arid public lands was carried on under the auspices of the Smithsonian Institution and, later, under an organization or bureau of the Department of the Interior known as the United States Geographical and Geological Survey of the Rocky Mountain Region. In 1879 this and other surveys maintained by the various departments of the Government were discontinued, and in their place a single bureau, the present United States Geological Survey, was created. Clarence King was the first Director, but he was succeeded after a few months by Major Powell, who then organized the work that laid the foundation of the great irrigation development which has transpired, and is to transpire upon a much larger scale in the future, in the Western States and Territories.

The first requirement of progress was exact information. It was necessary that the public lands should be classified, that the streams should be measured and mapped, that reservoir sites should be discovered and explored, and that the whole physical basis of the region, including its climate, should be reduced to a matter of scientific knowledge. This meant nothing less than the

THE CONQUEST OF ARID AMERICA

preparation of a complete topographical map, drawn in such detail that it should show all the elevations by means of contours, the location of streams, towns, roads, railroads, canals, isolated houses, and boundaries of States, counties, and towns. For a quarter of a century this work has gone forward until, at last, the secrets of the wilderness have been brought to light and the way prepared for its occupation by the hosts of civilization. The results are preserved in a number of annual reports covering what is popularly known as "the Powell Irrigation Survey."

In 1887 Congress recognized the importance of the investigation in its relation to practical progress and large appropriations were made for the study of specific reservoir sites. In 1889 a committee of the United States Senate, headed by Senator Stewart of Nevada, made a personal investigation of the arid region by means of an extended tour, and gave public hearings at numerous points.

Up to 1890 there was nothing which could be regarded as a public sentiment in support of irrigation as a broad economic movement and, much less, anything in the nature of organized public sentiment. Neither was there a popular literature to bring the matter to the attention of the masses. Irrigation was an unpleasant word, repellent and depressing. The word "arid" was synonymous with worthlessness. Scientific men like Major Powell, social reformers like Richard J. Hinton, and a few members of Congress who urged appropriations to assist Western development, were not taken seriously by the country at large. It was felt that they were ahead

THE RISE OF A NEW CAUSE

of their time and that at best there could be nothing but a sectional interest in the matter with which they dealt. There was no appreciation of the magnitude of the opportunity awaiting the Nation in the West, nor of the social and political significance of the work which must ultimately be done.

Such was the situation when the region of the Great Plains was overtaken by the drought of 1890, a calamity so deep and widespread that it staggered even the optimism of the West. While it was known and frankly acknowledged that irrigation was necessary in many localities west of the Rocky Mountains, the men of the semi-arid plains clung stubbornly to the belief that, in some mysterious manner, rainfall increased with railroad building, settlement, and the cultivation of the land. This delusion was effectually dispelled by the great drought, as related in a previous chapter. The psychological moment had come for the rise of a new cause which should take hold of the popular heart and go on, by a process of gradual unfoldment, until it became perhaps the greatest constructive movement of its time. Of this new and momentous epoch in Western history I am able to speak at first hand, since I can say, in the words of the ancient chronicler, "all of which I saw, and a part of which I was." *

* "Popular interest in irrigation was greatly stimulated by the discussion arising out of the Powell Irrigation Survey and the controversies over the report of the Senate Committee on Irrigation. This led finally to the holding of a series of national irrigation congresses, the prime mover in which was Mr. Wm. E. Smythe, of San Diego, Cal. The first was held in Salt Lake City, Utah, September 15 to 17, 1891."—*First Annual Report of the United States Reclamation Service.*

THE CONQUEST OF ARID AMERICA

In 1890 I was an editorial writer on the *Omaha Bee,* under that strong and able leader of Nebraska public opinion, Edward Rosewater. During the previous summer I had made a brief trip to the Maxwell land grant in New Mexico and for the first time saw men engaged in turning water upon land to make good the deficiencies of rainfall. I suppose I had heard or read the word "irrigation," though I have no recollection of it. Certainly, the word meant nothing to me until the drought struck Nebraska a year later. Then the thought occurred to me that the several fine streams flowing through the state might be employed to excellent advantage. Men were shooting their horses and abandoning their farms, within sight of these streams. There were the soil, the sunshine, and the waters, but the people did not understand the secret of prosperity, even with such broad hints before their eyes.

I thought of the thrifty orchards and gardens I had seen on the Las Animas and the Vermejo a few hundred miles farther southwest, and when Mr. Rosewater directed me to write editorials urging the public to contribute money, food, and seed for the drought-stricken farmers of Nebraska, I suggested that these should be supplemented by a series of papers dealing with the possibilities of irrigation. He gave me permission to do so on condition that I would sign the articles myself, as it was then considered little less than a libel to say that irrigation was needed in that part of the country.

How many lives those articles influenced, or are even yet to influence through the forces they set in motion, I do not know; but they changed my life completely. I

THE RISE OF A NEW CAUSE

had taken the cross of a new crusade. To my mind, irrigation seemed the biggest thing in the world. It was not merely a matter of ditches and acres, but a philosophy, a religion, and a programme of practical statesmanship rolled into one. There was apparently no such thing as ever getting to the bottom of the subject, for it expanded in all directions and grew in importance with each unfoldment. Of course, all this was not realized at first, yet from the beginning I was deeply impressed with the magnitude of the work that had fallen to my hand and knew that I must cut loose from all other interests and endeavor to rouse the Nation to a realizing sense of its duty and opportunity.

The first result of the articles in the *Bee* was a series of irrigation conventions in western Nebraska, beginning with the one at Culbertson, the seat of Hitchcock County. These county gatherings led to a state convention at Lincoln, and the state convention made me chairman of a committee to arrange for a National Irrigation Congress, which was held a few months later at Salt Lake, within sight of the historic ditch on City Creek where English-speaking men began the conquest of the desert.

I resigned my comfortable place on the *Bee,* launched The *Irrigation Age* (the first journal of its kind in the world, so far as I know), and went forth to do what I could. It was my rare good fortune to find a life-work, while yet on the sunny side of thirty, to which I could give my heart and soul with all a young man's enthusiasm.

It would be entirely erroneous to give the impression that the Irrigation Congress began its history by advocating the policy of improvement which finally prevailed. It was clearly realized that there was an enormous work to be done, and that this work could not be accomplished under laws then in existence nor by exclusive dependence upon private enterprise. But in 1891 there was no such sentiment for public ownership as now prevails. Speculation in water was considered as legitimate as speculation in land or mines. Moreover, it seemed scarcely conceivable that the Nation could be interested, within the lifetime of men then active in the West, to the extent of building reservoirs and canals and executing a great plan of colonizing arid lands. The most that was then hoped for was that Congress could be induced to cede the lands to the several States and Territories in which they were situated, and that when land and water were thus brought under one jurisdiction, laws could be devised to facilitate development. Each State would then be left to frame its own policy and the rivalry which would naturally ensue would inaugurate an era of tremendous activity.

This view was so generally entertained that Governor Arthur L. Thomas, in issuing his invitation for the meeting in Utah, set it forth in his formal call. The chief object of the convention was to consider the cession of the arid lands to the States. After several days discussion, in which the leading men of the West participated, the plan was approved without a dissenting vote, although J. W. Gregory, a delegate from Kansas, had pointed out certain perils inherent in the policy, in a speech which commanded attention.

The Salt Lake Congress precipitated a discussion

THE RISE OF A NEW CAUSE

throughout the country which resulted in the gradual growth of deep distrust in the plan of cession. It was strongly opposed by the leading newspapers of California, on the ground that all previous experience of the kind showed that the Western States would not deal wisely or honestly with such a grant. It was argued that the lands could not be utilized without a vast expenditure of capital, and that the inevitable way of obtaining this capital would be through the gift of the public lands to private interests, which would convert them into great estates. In a word, the movement was denounced as a gigantic scheme of land-grabbing, though it undoubtedly represented the best thought of the Western people at that time. It was seen that something must be done; it was perfectly obvious that the Nation would do nothing involving the expenditure of large sums from its own treasury; it was believed that if the States could obtain the land they would devise a means of preparing them for settlement. However, the opposition was strong enough to hold the movement in check and to create an interest in the discussion which was doubtless indispensable to progress of any kind.

The Second Irrigation Congress was held at Los Angeles, California, in 1893, and gained great distinction from its international character. Delegates were present from many foreign countries and the tone of discussion was entirely different from that at the Salt Lake meeting, two years previous. The keynote of the official pronouncement was that "the irrigation question is national in its essence." Lionel A. Sheldon, chairman of the committee on resolutions, aroused extraordinary enthusiasm when he declared his opinion that the arid lands would never be reclaimed until the Nation itself built the reservoirs and canals. But the convention wisely recognized that public opinion was not ready to support a specific proposal along this line, not even public opinion in the West. Consequently, it proceeded to organize commissions from its own membership in every State and Territory of the arid region, to canvass public sentiment and frame a plan for presentation at the next Congress.

The third meeting was held at Denver, Colorado, in 1894. The State commissions were unable to agree upon any comprehensive policy to be urged upon the lawmakers at Washington. But they advocated four steps of progress: first, the reform and unification of local water laws for the several States; second, the repeal of the desert land law; third, increased appropriations for the investigation of water supplies; fourth, the creation of a national commission to devise plans for the reclamation of arid lands. Subsequent sessions of the Irrigation Congress, held at Albuquerque, New Mexico, in 1895; at Phœnix, Arizona, in 1896; at Lincoln, Nebraska, in 1897; at Cheyenne, Wyoming, in 1898; and at Missoula, Montana, in 1899; followed the same lines, but with growing insistence on the national obligation to make the public domain fit for settlement by direct action of some kind.

While the cession movement inaugurated at Salt Lake had lost most of its force, Senator Warren, of Wyoming, introduced a bill to carry it into effect. It was not seriously considered by Congress. His colleague, Senator Carey, succeeded in passing a bill granting a million

A SAMPLE OF GOVERNMENT WORKS, NEVADA.

THE RISE OF A NEW CAUSE

acres to each of the States. Several accepted the gift and considerable development resulted from the policy, the common method being to grant to corporations the right to irrigate the land and dispose of it to settlers.

If the year 1879 is notable in irrigation history because of the publication of Major Powell's report on "The Lands of the Arid Region," and the year 1891 because of the organization of the National Irrigation Congress, the year 1897 is memorable because of new forces which came into it and exerted a powerful influence in shaping events. In the latter year Captain Hiram M. Chittenden, of the Corps of Engineers, U. S. A., published his report on "Reservoirs in the Arid Region." He had formerly been in charge of government works in the Yellowstone National Park and on important Western rivers. He had delved deep into the history of all the movements of population in the Far West and given much thought to the future civilization of the region. He looked upon the subject with the mind of a man trained in the government school of thought. Assigned to the study of reservoir problems on certain rivers of the West, he recommended that the Government should acquire full title and jurisdiction to any reservoir site which it might improve, and full right to the water necessary to fill the reservoir; also that it should build, own, and operate the works, holding the stored waters absolutely free for public use under local regulations.

The Chittenden report represented the break of day. Here was a clear suggestion of a workable plan, coming with the force of a recommendation from a distinguished engineer in the War Department. What was needed at this juncture was an organized propaganda, alive, tireless, sleepless. The Irrigation Congress had done a great work and years of usefulness were yet reserved to it. But it had no funds or paid officers. It met but once a year at widely separated points and always with a different membership. The time had now come when the cause required a working organism quite as effective as that of a church, a political party, or a great business enterprise.

This need was met by George H. Maxwell and his National Irrigation Association, the latter formed at Wichita, Kansas, in 1897, at the close of a meeting of the Trans-Mississippi Congress. Mr. Maxwell was an energetic young lawyer of California, with a remarkable talent for organization and a gift of forceful and eloquent speech. He was one of the numerous converts of the Irrigation Congress, which he joined at the Phœnix convention in the previous year. He determined to abandon his law practice and devote himself exclusively to the irrigation propaganda and the solution of other social problems which, as he clearly foresaw, must go hand-in-hand with the great scheme of reclaiming millions of acres of arid lands. The National Irrigation Association was not to supplant, but to strengthen and supplement, the pioneer institution, the National Irrigation Congress.

Mr. Maxwell saw that nothing could be done without a promotion fund. There must be offices in leading cities, periodicals and newspaper bureaus, and constant activity on the platform. Who should finance the great

THE RISE OF A NEW CAUSE

undertaking? Why not the numerous industrial and transportation interests, who would be the inevitable beneficiaries of new agricultural districts throughout the Western half of the continent and the resulting movement of people and products? Mr. Maxwell believed that if the managers of these enterprises understood their true interests, they would give liberal support to a work of this kind. He proceeded to convince them of the fact, and was thus enabled to carry on the propaganda with a vigor and success unprecedented in the history of the movement. He found an able and indefatigable lieutenant in Mr. C. B. Boothe, a prominent merchant of Los Angeles, California.

The Ninth Irrigation Congress assembled at Chicago in the autumn of 1900 and adopted ringing resolutions in favor of a comprehensive national system for the storage of floods and the reclamation of public lands. It demanded the abolition of water monopoly, insisting that water be made appurtenant to the land irrigated and that beneficial use be the basis, the measure, and the limit of the right.

The politicians were not slow to recognize the appearance of a new issue on the horizon. Thus in 1900, nine years after the first Irrigation Congress at Salt Lake and three years subsequent to the Chittenden report and the formation of the National Irrigation Association, the three great parties placed the following planks in their platforms:

Republican Platform.

In further pursuance of the constant policy of the Republican party to provide free homes on the public domain, we recommend adequate national legislation to reclaim the arid lands of

THE CONQUEST OF ARID AMERICA

the United States, reserving control of the distribution of water for irrigation to the respective States and Territories.

Democratic Platform.

We favor an intelligent system of improving the arid lands of the West, storing the waters for the purposes of irrigation, and the holding of such lands for actual settlers.

Silver Republican Platform.

We believe the National Government should lend every aid, encouragement, and assistance toward the reclamation of the arid lands of the United States, and to that end we are in favor of a comprehensive survey thereof, and an immediate ascertainment of the water supply available for such reclamation, and we believe it to be the duty of the general government to provide for the construction of storage reservoirs and irrigation works, so that the water supply of the arid region may be utilized to the greatest possible extent in the interests of the people while preserving all rights of the State.

The first stage of the battle had been won. Irrigation was squarely before the American people as a question which must be dealt with. It was no longer merely the dream of enthusiasts, but a subject which thundered at the door of Congress and demanded the attention of practical statesmanship.

CHAPTER II

ON THE ANVIL OF CONGRESS

FOR many years prior to the indorsement of a national irrigation policy by the great political parties, Western Senators and Representatives had been in the habit of introducing bills aiming at the reclamation of arid lands in their own States or districts. These measures were fruitless, not only because the East had not yet given its assent to this form of internal improvement, but because it was impossible to unite the West in favor of appropriations for any particular locality. Each neighborhood stood ready to cheer "for the old flag and an appropriation," but could evoke no enthusiasm when it was proposed to spend the appropriation in some other neighborhood.

The West had not learned its lesson thoroughly when the Fifty-sixth Congress assembled for its final session, on December 3, 1900. It had not devised a way to overcome Eastern and Southern opposition to direct appropriations nor to unite its own members on a single, comprehensive plan of development. President McKinley made no mention of irrigation in his message. On the very first day of the session, Representative John F. Shafroth of Colorado introduced a bill calling for an appropriation of thirteen million dollars for the reclamation of arid lands. It was not a local bill, but it proposed to take the money straight from the national treasury, conceding nothing to the prejudices of the East or the South on that subject. There was, therefore, no hope of its success. Representatives Newlands of Nevada and Bell of Colorado followed with bills aiming at peculiar benefits for their own States. These measures were impossible, because the West would not unite on them.

On January 26, 1901, the national irrigation cause saw its first real daylight in the halls of Congress. It was upon that memorable date that Francis G. Newlands, then Representative and now Senator, from Nevada, introduced the first of a series of measures, each an improvement upon its predecessor, but all based upon the same fundamental principles and all widely discussed by press, public, and law-makers, under the general title of "the Newlands bill." A comparison of the original measure of January 26 with the present law as it stands upon the statute books amply justifies the use of the Nevadan's name as the real author of the present successful policy.*

The Newlands Bill proposed a continuing appropriation to be derived from the sale of public lands, and put this appropriation at the disposal of the Secretary of the Interior for use not only in making investigations, but for actual construction of reservoirs and canals. This

* The original "Newlands bill" of January 26, 1901, together with the full text of the famous Act of June 17, 1902, are published in the Appendix of this volume, for the benefit of those who desire to put this statement to the test of critical analysis.

FRANCIS G. NEWLANDS.—Whose famous irrigation bill became the foundation of a New National Policy.

ON THE ANVIL OF CONGRESS

was the foundation principle of the law enacted on June 17, 1902.

None save those quite familiar with the difficulties which must be surmounted in order to obtain the necessary support to pass any irrigation law, and then to have it workable in the highest degree without constant attention from Congress in the future, can possibly appreciate how remarkably this measure was adapted to the situation. It solved at a single stroke questions which could not have been solved in any other way except by years of effort and, probably, a considerable body of legislation.

First of all, the plan disposed of the Eastern objection to direct appropriations from the treasury. Money received from the sale of public lands came exclusively from the West and the major portion of it from the arid region. It was money paid by homeseekers. Why should it not be expended in the West and for the purpose of making homes? Certainly, it did not represent taxes paid by Eastern people. The argument against creating new competition for Eastern farmers could not stand alone. It must be supported by the further argument that it was unfair to tax the East for the specific purpose of creating such competition. No one was radical enough to say that there should be no new homes or farms in the United States and that the historic homestead policy must be utterly abandoned. This being so, the fiscal feature of the Newlands plan could not be successfully attacked on either logical or patriotic grounds. In a word, it supplied the unanswerable solution of the most troublesome feature of the problem.

Few realized at the time how completely the continuing

appropriation, derived from the receipts for land sales, removed the question from the sphere of congressional action. It put at the disposal of the Secretary of the Interior a large and growing fund, to be expended and collected and used over and over, for an indefinite period in the future. It was no longer necessary to go to Congress and ask for the approval of a specific project, and for the money to build it. The money was provided in advance, while the Secretary had full power to make the investigations, approve a project, set apart the necessary funds, and proceed with its construction. So far as Congress was concerned, the measure was the most remarkable piece of automatic legislation ever devised. In subsequent pages we shall see how well it has worked in practical operation and what weary years have been saved by the concentration of such authority in the hands of the Secretary of the Interior.

The plan also supplied the only satisfactory solution of the questions arising from local jealousies in the West. The money was available for irrigation in sixteen States and Territories. Investigations were to be made in all of them and the Secretary of the Interior was empowered to determine where it was desirable to make improvements first. It was assumed that he would consider the claims of all localities and arrive at his conclusions from the broad national standpoint. This expectation has been fully met by the manner in which Secretary Hitchcock has administered the law. It is hardly to be believed that equally good results would have emerged from the hurly-burly of Congress if the matter had been handled after the manner of a river and harbor bill.

GOVERNMENT ROAD-BUILDING NEAR ROOSEVELT DAM, ARIZONA.

ON THE ANVIL OF CONGRESS

Other important features of the measure which were later embodied in actual legislation, were these: provision for the withdrawal from entry, in the discretion of the Secretary of the Interior, of all public lands required for reservoirs or canals, or susceptible of irrigation from proposed works; pro-rating the cost among lands irrigated and providing for the repayment of the amount to the reclamation fund in ten annual instalments; making the water right perpetually appurtenant to the land, with beneficial use the basis, the measure, and the limit of the right; providing that land could be taken only under the homestead law, with its requirement for actual settlement; and permitting the sale of water rights to land in private ownership, but only in small tracts.

The virtues of the bill were promptly recognized by those most familiar with the needs of the West. Mr. Frederick H. Newell, then Chief Hydrographer of the Geological Survey and now Chief Engineer of the Reclamation Service, testified before a Committee of Congress: "Mr. Newlands's general bill has been so worded as to avoid striking on all the snags which are impeding the progress of the development and reclamation of the arid lands." Mr. George H. Maxwell, Executive Chairman of the National Irrigation Association, testified:

" I wish to speak of the Newlands Bill, No. 14,088. I think a good name for it would be ' the omnibus bill.' It is undoubtedly true that after fifteen or twenty years of Government investigation we are no further than we were at the beginning, so far as the actual reclamation of the land is concerned. Under this bill the Government can begin construction immediately and I believe along lines which remove every reasonable objection to the Government undertaking the great work of bringing about

THE CONQUEST OF ARID AMERICA

the reclamation of the arid lands. The first point which it seems to me is important in favor of the Newlands Bill is that under it everything can be done which is suggested to be done by each of the other bills now before this Committee."

Mr. Newlands called a conference at his home in Washington of seventeen Senators and Representatives from the Arid States, without regard to their party affiliations. This conference approved the measure and it was introduced in the Senate by Senator Hansbrough, of North Dakota, January 30, 1901, and reported favorably by the Senate Committee on Public Lands five days later. About the same time, it was also approved by the House Committee on Irrigation. The measure provoked an animated debate in the House of Representatives, but did not reach the voting stage, as the short session expired by limitation on the fourth of March.

The results attained in Congress in the brief session of three months, were very remarkable. A measure had been framed which effectually disposed of the most serious objections to a national irrigation policy entertained in the East, and this same measure completely overcame the rivalries of Western communities, each of which sincerely believed it presented the best opportunity for the initial national enterprise. The plan won the approval of the organized irrigation movement and the approval of the country. Victory seemed almost in sight when the Fifty-sixth Congress adjourned, at the close of William McKinley's first administration.

CHAPTER III

IRRIGATION IN THE WHITE HOUSE

THEODORE ROOSEVELT succeeded to the Presidency on September 15th, 1901. He was the first occupant of the White House who had seen enough of the Far West to comprehend its unique economic possibilities and to understand its claims upon the attention of the Nation.

The national irrigation movement was born under the administration of Benjamin Harrison. The Indiana President had, indeed, made an extended tour of the arid region during the year that the first Irrigation Congress was held at Salt Lake. In the course of his remarkable speeches he paid frequent tribute to the hardihood and enterprise of the pioneers who turned the streams from their channels and so made oases in the desert. But it was not given to him to behold the true significance of irrigation. And it was with genuine relief that, on his return to the humid region, he looked out on green fields and remarked to an early morning audience, " It is good to be back where God furnishes the rain."

Ten years elapsed before another President beheld the same scenes in his progress to the Pacific. During the interval, the cause of reclamation had made a wonderful advance in popular estimation, yet William McKinley,

THE CONQUEST OF ARID AMERICA

who was singularly gifted with the power to discern what the people were thinking about, did not realize that he was in the presence of one of the great issues of the imminent future. To an intimate friend he remarked: "I can see that sometime there will be a vast problem here for the people to solve, but it will come long after I have passed from the stage of events." Within six months of that time a presidential message gave large attention to the subject and urged immediate action—but it was written by another hand. And within less than one year a national irrigation law had been enacted—but William McKinley was in his honored grave.

When a young man, in somewhat delicate health, Mr. Roosevelt came out from the East to seek the strength of the mountains and the benediction of the unclouded sun. He made a ranch on the headwaters of the Little Missouri, herded cattle, mingled with cowboys, hunted the big game of the plains. There he learned the marvel of the arid soil when joined to the waters of the mountain stream. And there he became essentially a Western man in spirit and in temperament. It would have been strange if, in his long exile in the unpeopled wilderness, he had not pondered upon the ultimate future of the region. It was thus natural enough that he should take kindly to the idea of national irrigation when it had been brought prominently to the attention of the country by its aggressive champions; and his warm letter of commendation addressed to the National Irrigation Congress at Chicago in 1900, when he was Governor of New York, surprised no one who knew his partiality for the West. The tragic death of President McKinley brought to power

From stereograph. Copyright, 1905, by Underwood & Underwood, New York.

THEODORE ROOSEVELT, THE IRRIGATION PRESIDENT.

IRRIGATION IN THE WHITE HOUSE

the one American citizen then regarded as a Presidential possibility who was thoroughly committed to the national irrigation plan. Upon entering the White House, he at once declared his intention to press the new policy, and proceeded to use all the influence of his great office in urging it upon the attention of Congress and the country.

President Roosevelt's first message, delivered December 3d, 1901, outlined a much broader basis for national irrigation than most of its advocates had considered practicable up to that time. It went the full length of the Chittenden recommendations and was easily susceptible of a construction which carried it even farther. To illustrate, its opening paragraphs did not deal with the question of the public domain, primarily, but with the wider and deeper question of laying a foundation for the economic life of the West. This, of course, comprehends the whole physical basis of the region, including property now in private as well as public ownership. The President saw no reason why a scheme of internal improvements in the arid region should be discussed in its relation to individual and local benefits, as distinguished from the benefits to be conferred upon the Nation as a whole, any more than individual and local benefits are discussed in relation to improvements in the Mississippi Valley or on the Atlantic Coast. He put the new policy on precisely the same basis as the old. To his mind, both represented an exercise of national power for the protection and development of national resources. And one was as necessary and legitimate as the other. Here are the words of his epoch-making recommendations:

THE CONQUEST OF ARID AMERICA

"The forests alone cannot, however, fully regulate and conserve the waters of the arid region. Great storage works are necessary to equalize the flow of streams and to save the flood waters. Their construction has been conclusively shown to be an undertaking too vast for private effort. Nor can it be best accomplished by the individual States acting alone. Far-reaching interstate problems are involved ; and the resources of single States would often be inadequate.

"It is properly a national function, at least in some of its features. It is as right for the National Government to make the streams and rivers of the arid region useful by engineering works for water storage as to make useful the rivers and harbors of the humid region by engineering works of another kind. The storing of the floods in reservoirs at the headwaters of our rivers is but an enlargement of our present policy of river control, under which levees are built on the lower reaches of the same streams.

"The Government should construct and maintain these reservoirs as it does other public works. Where their purpose is to regulate the flow of streams, the water should be turned freely into the channels in the dry season to take the same course under the same laws as the natural flow."

It was thus that the President invited the Nation to enter upon a new and magnificent enterprise. And it was only when he had done so that the friends of the policy realized how well they had done their work of popular education. The idea of conquering half a continent for civilization flattered the national pride and appealed irresistibly to the national imagination. Not only was this true of the West, where the Presidential message fell like the stroke of high noon on the clock of destiny, but of the Middle States, the South, and the North Atlantic.

That portion of the message dealing with the reclamation of the public domain was only second in importance

IRRIGATION IN THE WHITE HOUSE

to what the President had said of the larger aspects of the question. It was as follows:

" The reclamation of the unsettled arid public lands presents a different problem. Here it is not enough to regulate the flow of streams. The object of the Government is to dispose of the land to settlers who will build homes upon it. To accomplish this object water must be brought within their reach.

" The pioneer settlers on the arid public domain chose their homes along streams from which they could themselves divert the water to reclaim their holdings. Such opportunities are practically gone. There remain, however, vast areas of public land which can be made available for homestead settlement, but only by reservoirs and main-line canals impracticable for private enterprise. These irrigation works should be built by the National Government. The lands reclaimed by them should be reserved by the Government for actual settlers, and the cost of construction should, so far as possible, be repaid by the land reclaimed. The distribution of the water, the division of the streams among irrigators, should be left to the settlers themselves, in conformity with State laws and without interference with those laws or with vested rights."

The irrigation measure on which the West had practically agreed was again introduced on the day the Fifty-seventh Congress assembled for its first session,—December 2, 1901,—by Representative Newlands, and two days later by Senator Hansbrough. Then came the President's message with its inspiring assurance of unstinted executive support for the movement. The Western Senators and Representatives were again summoned into conference to prepare for the final fight in Congress. Attempts were made to alter the character of the measure materially, but they were unsuccessful with a single exception. The feature of the "original Newlands bill" providing for the withdrawal of lands from entry under all laws except the homestead, without the benefit of the commutation clause, until the works should be finished and the water actually ready for delivery, was stricken out by the Committee.

This action was intensely disappointing to the organized irrigation movement, who believed it was done solely in the interest of land-grabbers who desired to get possession of the choicest morsels of the public domain in advance of homeseekers. George H. Maxwell declared that it amounted to a betrayal of the most sacred objects of the movement and said it was infinitely preferable that the entire measure should be lost at that time rather than that a condition should be created under which it would be not only possible, but probable, that the lands would be stolen before the genuine homemaker could get an opportunity to file upon them. His aggressive stand aroused a storm of opposition to the amendment. It resulted in an animated conference at the White House, at which the President announced that he would not sign the bill in that shape. The original provision was then restored so that it was made impossible for any one to obtain title to public lands irrigated by the Government, without five years' residence and actual cultivation.

The Newlands' bill, which had previously passed the Senate, went through the House on June 13, 1902, by a vote of 146 to 55. It was signed by the President on June 17, the 127th anniversary of the battle of Bunker Hill.

The National Irrigation Law had a singular experi-

IRRIGATION IN THE WHITE HOUSE

ence in the presidential campaign of 1904. It is perhaps the only measure in the history of American legislation enjoying a popularity so absolute and unquestionable that the only possible controversy between the two parties was as to which was entitled to the greater credit for bringing it to pass. The controversy on this point raged fiercely throughout the arid region and was sharply accentuated by the Democratic demand for a policy of domestic development as opposed to foreign expansion.

The Republicans asserted that the President was chiefly entitled to the credit for the passage of the law, and claimed that the seventeen Senators and Representatives from the Arid States who united in recommending the measure were the real authors of the bill. The Democrats replied that the measure was framed and introduced by a statesman of their own faith, that it was passed in the Senate by a non-partisan vote, and passed in the House (where the real battle was fought) by a vote the majority of which was Democratic, over the opposition of the strongest Republican leaders in that body.

The controversy was one of more than passing interest because it involved the attitude of individuals and of parties in connection with principles which are certain to be much debated in the future and to exert a far-reaching influence on the course of events. It was hopeless to expect a judicial consideration of the matter in the height of a presidential campaign. Since then, however, it has been discussed in a spirit which seems entirely worthy of permanent record. At Sheridan, Wyoming, in July, 1905, in the presence of several of his colleagues on the Committee of Seventeen who urged the measure

THE CONQUEST OF ARID AMERICA

upon Congress, Senator Newlands reviewed the controversy with the utmost frankness. His presentation was received with a degree of enthusiasm and apparent unanimity which warrants the hope that the contention is now ended. The occasion was furnished by a banquet tendered to the members of the Congressional Committees on Irrigation, near the end of their long tour of the arid region. Senator Newlands said:

" I have listened through our journey to the praise bestowed upon President Roosevelt in connection with the irrigation movement, and I would not detract at all from the deserved reputation which he enjoys by his prominence in it. But I wish to call your attention to the fact that it was a union of Democrats with Mr. Roosevelt and Mr. Mondell and Mr. Jones of Washington and Mr. Reeder of Kansas and other Republicans like them, that accomplished this act of legislation."

Mr. Mondell: " You are right."

Mr. Newlands: " The fact is that we Democrats have adopted Mr. Roosevelt. And there are now two wings to the Democratic party, the Republican wing and the old Democratic wing. We are for Mr. Roosevelt's Democratic policy of reform, and we are going to see that that policy is enacted into law within the next three years, for Providence has assigned to him the opportunity of achievement."

Mr. Newlands then referred to the fact that throughout their pilgrimage in the Southwest and West, Mr. Mondell had referred to the famous Committee of Seventeen—a voluntary committee, selected, after Mr. Roosevelt became President, by the Senators and Representatives from the Western States, to adjust the differences of the West regarding irrigation, as the source and origin of the Reclamation Act.

GOVERNMENT CEMENT MILL, SALT RIVER PROJECT, ARIZONA.

IRRIGATION IN THE WHITE HOUSE

" This," he said, " was a mistake. The Reclamation Act, in all its essential features, had been framed and presented by a Democrat at the preceding session of Congress, immediately following the election of 1900 and whilst Mr. McKinley was living."

Referring to its history, Mr. Newlands stated that in 1900, after many years of agitation, the Western men had secured from both the Democratic and Republican parties a declaration in their platform favoring national irrigation, and at the session of Congress next ensuing (the session constituting the last of McKinley's first term), it was determined to press the matter. At that time, said Mr. Newlands, there were but three bills pending in Congress providing for immediate construction, which had been presented by Shafroth, Bell, and himself.

" As the Committee hearing progressed," said Mr. Newlands, " it soon developed that members differed greatly, and I came to the conclusion that we could not hope to persuade the East until the men of the West were united. And so, with a view to shaping a broad and comprehensive national measure that would receive the support of and include the entire arid region, I made a careful study of all previous bills, including those of Mr. Shafroth and Mr. Bell, from which most valuable suggestions were received. I also consulted Mr. Newell of the Reclamation Service, Mr. Maxwell of the Irrigation Association, Mr. Elwood Mead, and other irrigation experts, who differed widely as to the form of legislation; and finally, on the 26th day of January, 1901, I introduced in the House a bill which contained every essential feature of the Reclamation Act that is now upon the statute book.

" This bill provided for a revolving reclamation fund derived from the sales of public lands; it authorized the withdrawal from entry of lands necessary for irrigation projects, and provided for the immediate construction of irrigation works wherever deemed feasible by the Secretary of the Interior, upon whom the only restraint imposed was that no contract should be let unless the money for its payment was in the fund. It discouraged land monopoly by dividing the lands to be irrigated into small holdings for actual settlers, and sought to destroy existing land monopoly by providing that no private owner of land could secure a water right for more than eighty acres, thus compelling the division of large holdings into small farms. It provided for the payment of the cost of each project by the settlers, without interest, in ten annual instalments, the fund being thus retained as a revolving fund for future operations.

"This bill was considered at a meeting of Western Senators and Representatives at my house, and, upon motion of Senator Pettigrew, the bill was approved and Senator Hansbrough, a most strenuous advocate of national irrigation, was requested to introduce it in the Senate, which was done the next day. This bill was immediately accepted by the Western newspapers and Western sentiment as a satisfactory solution of the question, and within six weeks and before the close of the session and of McKinley's first adminstration, the movement for its passage had made such headway that it received the sanction of the Senate Committee on Public Lands of the Senate, and its leading provisions had received the approval of the House Committee on Irrigation.

"All this was accomplished before Mr. Roosevelt became President. The Western press announced that the bill would be pressed by the West at the following session. Meanwhile, however, a movement was organized in Wyoming to defeat this bill and to substitute for it a measure more in harmony with the Wyoming view, which sought to retain State control over the construction and administration of irrigation projects. And so a convention was called at Cheyenne of Representatives from North and South Dakota, Nebraska, Kansas, Colorado, Wyoming, Montana, and Idaho, most of which States had irrigation departments organized under State Engineers in

IRRIGATION IN THE WHITE HOUSE

sympathy with the views of Wyoming. At this convention, of which Mr. Mondell, Senator Warren, and others were conspicuous members, the so-called 'State Engineers' Bill' was approved, a bill which accepted the provision of the general bill introduced at the preceding session of Congress by myself as to the creation of a reclamation fund, but provided for the construction and control of irrigation projects by the State Engineers of the respective States.

"Later on, when we met at Washington for the first session under Roosevelt's administration, a call was issued to the Senators and Representatives to meet at Senator Warren's committee room, and there Senator Warren presented the State Engineers' bill for our consideration. We who believed in thoroughly nationalizing irrigation, who believed that every river and its tributaries, regardless of State lines, should be made the subject of comprehensive study by the National Government with a view to the construction of works that would secure the largest development of the entire drainage area, vigorously fought this bill, and it was rejected.

"Then it was that the Committee of Seventeen, composed of one Senator or Representative from each State or Territory affected, was selected to harmonize the differences of the West. For thirty days this committee sat in session, their contention being mainly between those of us who wished thoroughly nationalized irrigation and those who wished to retain some form of State control; and at the end of that period of contention the Reclamation Act was reported containing every essential provision of the bill to which I have referred, which was introduced by me in the preceding Congress, and Mr. Hansbrough was instructed to offer this bill in the Senate and I was instructed to offer it in the House.

"The passage of this bill in the Senate was assured because of the large representation of the West in that body. The difficulty was in the House, where the Western delegation was proportionately small and in which the Republican leaders had arrayed themselves in opposition to the bill. The bill had to go before the regular committees for consideration. It was reported to the House by Mr. Mondell, who had accepted the judgment of the West as to the complete nationalization of irrigation, in a very comprehensive report. It was at this time that President Roosevelt, whose message had drawn the attention of the entire country to the importance of irrigation, intervened as the friend of the West and gave his powerful influence to breaking down the opposition of Republican leaders. It was largely due to him that consideration of the bill was conceded by the House leaders who, although they voted against the bill, relaxed their opposition to its consideration."

Supplementing an account given by Mr. Mondell of an interview with Mr. Roosevelt, in which Mr. Roosevelt gave him a letter addressed to Mr. Cannon urging the immediate consideration and passage of the bill, Mr. Newlands referred to an interview of his own with the President in which the latter expressed the greatest interest in the subject, but suggested that it might be better to substitute for the pending comprehensive bill a bill for a small project involving the expenditure of $250,000 or $500,000, as an entering wedge for future legislation, Mr. Newlands continued:

"My reply was: 'Mr. President, we can pass a big bill as easily as we can pass a little bill. If we pass a little bill for one project, it will take five years for its completion, and Congress will then take five years longer in determining whether it is a success or not; besides, we cannot stake the cause of national irrigation on the success of one project, which might be a failure. We can pass this general bill if you can moderate the opposition of the leading members of your own party, for the Democratic party stands ready to support this bill.' It was not necessary to urge this upon President Roosevelt, who was heart and soul in the movement, and who threw himself into the advocacy of the movement with a zeal all his own. The result was that while the leaders to whom I have referred voted against the bill, they graciously allowed it to be considered in

IRRIGATION IN THE WHITE HOUSE

the House, and after a spirited debate, in which Mr. Shafroth took so conspicuous a part, the bill was passed.

"Let me say that though the House was Republican, the majority of the votes which passed the bill was Democratic; and of the votes against the bill, three-fourths were Republican and one-fourth Democratic. And so I say that whilst we recognize the splendid advocacy of Roosevelt and the energetic efforts of other Republicans like him, I insist upon it that history records that the bill in all its essential features was framed by a Democrat and passed in the House of Representatives by a vote the majority of which was Democratic; and that the bill stands upon the statute book to-day as the result of a union of Democracy with President Roosevelt and his Republican friends of like faith."

Mr. Newlands then went on to declare that the Democrats would stand with the President in the reform of the land laws and other measures. He commented at length upon the necessity of reforming the land laws, insisting that through their operation the natural wealth of the country, the wealth of timber, of coal, of iron, and of oil in the public domain belonging to the entire people, was drifting into the hands of syndicates and monopolies, to be used for their oppression.

"I insist," said Mr. Newlands, "that these questions are too serious for mere partisan consideration. The democracy of both parties should be aroused to the necessity of upholding President Roosevelt in his domestic policies. No desire to embroil the President with his own party should control the sentiment of Democrats, but rather the patriotic motive of aiding him to place upon the statute books the policies of which he is to-day the leading exponent. A greater democracy should be appealed to—not the mere democracy of a party, but that larger democracy which enrolls as its distinguished leaders Jefferson and Jackson, Lincoln and Roosevelt."

CHAPTER IV

UNCLE SAM'S YOUNG MEN AT WORK

THE main provisions of the national irrigation law are as follows:

Beginning with the fiscal year ending June 30, 1901, the entire receipts from the sale of public lands in Arizona, California, Colorado, Idaho, Kansas, Montana, Nebraska, Nevada, New Mexico, North Dakota, Oklahoma, Oregon, South Dakota, Utah, Washington, and Wyoming, are set aside and appropriated as a special fund in the Treasury, to be known as the Reclamation Fund, for the examination and survey, and for the construction and maintenance, of irrigation works for the storage, diversion, and development of waters for the reclamation of arid and semi-arid lands in those States and Territories.

The Secretary of the Interior is authorized and directed to make examinations and surveys for, and to locate and construct, irrigation works for the storage, diversion, and development of waters, including artesian wells, and to report to Congress all the details of the work accomplished.

The Secretary is required, before giving the public notice of his plans, to withdraw from public entry both the lands which may be needed for irrigation works and the lands which are likely to be reclaimed by such works, except that agricultural lands shall be open to entry under the homestead law, but without the privilege of the commutation clause of that law.

The Secretary has full authority to let contracts for the construction of such irrigation works as he considers feasible, providing the necessary funds are available. He is not required to wait for any special authorization by Congress.

The working day for men employed in construction is fixed at eight hours and the employment of Mongolian labor is prohibited.

The Secretary determines the amount of land which each settler may file upon, but the law fixes the minimum entry at 40 acres and the maximum at 160 acres. Within these limitations, discretion is left to the Secretary and he is expected to determine the area "reasonably required for the support of a family upon the lands in question."

The law distinctly contemplates the reclamation of lands in private ownership and makes it the duty of the Secretary to determine the terms upon which water shall be supplied to such lands, provided "no right to the use of water for land in private ownership shall be sold for a tract exceeding 160 acres to any one landowner, and no such sale shall be made to any landowner unless he be an actual bona fide resident on such land, or occupant thereof, residing in the neighborhood." (The last clause was added for the accommodation of settlers who, like the Mormons, prefer to have their homes in village centres).

Settlers are required to pay the usual Government price of $1.25 per acre for land, and, in addition, the price fixed by the Secretary for water. The latter is sufficient to reimburse the Government for the cost of the works and to be paid in ten annual instalments, without interest. This applies alike to private landowners and to entrymen on public land.

When payment has been made for the major portion of the lands irrigated by a given system, the management and operation of the irrigation works passes to the landowners under such form of organization and such rules and regulations as the Secretary may approve.

But the title to reservoirs and to the works necessary for their protection and operation, together with the management of same, rests with the Government until otherwise provided by Congress.

The Secretary is given authority to acquire by purchase or condemnation any rights or property which he finds necessary in the application of the law.

The Secretary is required to expend the major portion of the funds arising from the sale of public lands within the State or Territory whence the money is derived, but may temporarily use the fund wherever he may deem advisable. At least once in ten years, the expenditures for the benefit of the various States and Territories must be equalized, as far as this may be feasible.

Nothing in the law is intended to affect or interfere with local statutes relating to the control, appropriation, or distribution of water, or with vested rights. It is provided, however, that the right to the use of water from the national system shall always be appurtenant to the land, and that beneficial use shall be recognized as the basis, the measure, and the limit of the right.

ETHAN ALLEN HITCHCOCK, SECRETARY OF THE INTERIOR.—Whose name will live in History as the First Administrator of National Irrigation and Fearless Prosecutor of Land Frauds.

UNCLE SAM'S YOUNG MEN AT WORK

Within twenty-four hours after the signing of the measure by the President, Secretary Hitchcock had set the machinery of the Department of the Interior in motion to carry it into effect. He assigned the details of administration to the Geological Survey. Director Charles D. Walcott immediately organized the United States Reclamation Service, putting at the head of it a trained man of great competence, Frederick Haynes Newell.

Educated at the Massachusetts Institute of Technology, Chief Engineer Newell had already spent a dozen busy years in studying the hydraulic problems of the arid region. He was one of the young men who had grown up in the inspiring presence of John Wesley Powell, founder of the Geological Survey, and first scientific explorer of the Far West. Major Powell had predicted for him a great career and had done much to fit him for the work upon which he now entered. He came to his new task a patient, thorough, scientific servant of the Republic, with no dangerous enthusiasms, but with deeply-founded faith in the value of the great enterprise committed to his hands.

The work of the Reclamation Service is planned by a Board of Consulting Engineers, composed of men who represent the finest talent and most valuable experience for this particular service in the United States. These men are Arthur P. Davis, Joseph Barlow Lippincott, G. Y. Wisner, H. N. Savage, J. H. Quinton, W. H. Sanders, and Benjamin M. Hall. They were all carefully trained for their profession, and they have all been associated with the work of designing and constructing

THE CONQUEST OF ARID AMERICA

important public and private hydraulic systems in many different parts of the country. In April, 1905, the President appointed C. E. Grunsky, of California, late a member of the Isthmian Canal Commission, Consulting Engineer to the Director of the Geological Survey, which further strengthened the Reclamation Service, as the Director stands between Chief Engineer Newell and the Secretary of the Interior in the administration of the great enterprise.

The first work of the Reclamation Service was to organize a corps of young men to make the necessary surveys and examinations over a vast territory. The nucleus of the service consists of men who have been engaged in measuring streams and studying the economic possibilities of the West under the Geological Survey. Additional men are obtained through competitive examination. These are of two classes. The first consists of experienced men chosen through examinations based on practical questions where the rating depends largely on the applicant's record in the construction of irrigation works. The second is made up of well-educated young men of good character who enter the lower ranks with the expectation of being advanced as opportunity offers. The young fellows fresh from college receive while on probation about $60 per month; as they show their worth, they are advanced to $75 per month, then to $1000 a year. Up to this stage they are known as engineering aids. Later, they become assistant engineers, receiving $1,200, $1,400, or $1,600 a year, until the time when they have demonstrated their ability to conduct independent work and to initiate plans. Then they become

1. CHARLES D. WALCOTT.—Director of the United States Geological Survey.
2. FREDERICK H. NEWELL.—Chief Engineer of the United States Reclamation Service.
3. GIFFORD PINCHOT.—Forester.
4. C. E. GRUNSKY.—Consulting Engineer to Director of the Geological Survey.

full engineers, receiving from $1,800 to $3,600, according to age and experience. Appointments and promotions are made in accordance with the best ideal of civil service and political influence is wholly ignored.

There are thirteen States and three Territories which constitute the field of operations for the Reclamation Service. In all of these, engineering parties were promptly put at work. Some of the parties were quite large. For instance, over one hundred men were engaged for several months on the lower reaches of the Colorado River in Arizona and California. Camps are frequently located in the wildest spots still remaining in the United States.

The first step everywhere is to measure streams and learn the quantity of water available for the reclamation of new areas. The next step, roughly to survey the lands susceptible of irrigation from the source of supply. When a project looks promising, the engineers proceed to obtain exact information concerning the cost and efficiency of necessary works, then to study all the factors entering into the economic questions presented. The field work completed, full reports concerning the project, together with maps and photographs, are forwarded to the Chief Engineer at Washington. This official gives the matter careful consideration and, if the facts are apparently complete, submits them to a consulting board of three or more engineers whose professional standing is such as to give confidence in their opinions. If the matter still looks promising, the board visits the locality, goes over the items, verifies the conclusions, and transmits its findings to the Chief Engineer. If additional surveys are needed, they are made; otherwise, the conclusions are sent forward with recommendations, through the Director of the Geological Survey, to the Secretary of the Interior.

The plans are not yet public property, and may never become such. All that has been done up to this point is for the information of the Administration, which must now exercise its own judgment as to how, when, and where the reclamation fund shall be expended. It is only when final action has been determined upon that the public may know what the Government will do.

Before the National Irrigation Law had been on the statute books a year, examinations had been pushed into nearly all States and Territories mentioned in the Act, and five projects had been officially approved. These were located, repectively, on the Milk River in Montana, the Sweetwater in Wyoming, the Truckee in Nevada, the Gunnison in Colorado, and the Salt River in Arizona. In another six months construction had actually begun in Nevada and Arizona. The amount in the Reclamation Fund on June 30th, 1903, the end of the second fiscal year, was about $16,000,000. At the end of the third year it was $23,000,000 and on June 30th, 1905, about $30,000,000.

There is no other instance in American history where the inauguration of a new policy of internal development was so promptly accomplished. Not an hour was wasted in useless talk, in Congress or out of it. In the fulness of time, Uncle Sam had laid his hand on the door of Arid America, and the whole Nation looked on with enthusiastic approval.

UNCLE SAM'S YOUNG MEN AT WORK

The Agricultural Department co-operates with the Reclamation Service in preparing the way for the settler. There is much to be done besides bringing water to the desert. The soil must be analyzed and its capacity for the production of various crops determined in advance, as far as possible. Various social and economic problems closely related to irrigation must be subjected to careful study. The duty of water—that is to say, the amount required to obtain the best results with certain crops and soils—is a matter of vital importance to the settler and only to be ascertained by scientific investigation.

The division of Irrigation Investigations and Drainage in the Bureau of Experiment Stations is one of the most important agencies now engaged in working out the problem of colonization. Elwood Mead, who had done one lifetime's work in framing and administering the water laws of Wyoming, is now engaged in the performance of another as Chief of this Division. He approached these problems with a grasp and a trained intelligence which no other man could have given them and, with the aid of his numerous staff of skilled and earnest assistants, he is guiding the work of the pioneer settlers in a way which will enable them to build on scientific foundations. While this work is less dramatic than the construction of reservoirs and canals, and far less monumental in its physical results, it is none the less important to the people of the country. It is, in fact, the work which will enable the future millions to utilize the opportunities opened by the Reclamation Service and to preserve these advantages to the latest generation.

CHAPTER V

PREPARING HOMES FOR THE PEOPLE

ON July 1st, 1905, the reclamation fund for home-making in Arid America amounted to about $30,000,000. Plans had been made for the expenditure of this amount in thirteen States and three Territories. On the 17th of June, the third anniversary of the Reclamation Act, the formal opening of the first government project occurred in Nevada in the presence of a large and distinguished company. Work is being vigorously prosecuted on many other projects, and it is expected that on each succeeding anniversary of the Act at least one system will be completed and a large area of lands dedicated to settlement.

In preceding chapters, the resources of various States and Territories have been discussed and the story of their development brought down to date. There the reader will find general descriptions of climate, soil, geography, markets, and other considerations of vital interest to the intending settler in all localities where works of irrigation are building or to be built. The object of this chapter is to supply concise and reliable information concerning projects actually under way and certain to be opened to homeseekers in the early future.

Any unmarried person over twenty-one years of age, or head of a family, who is, or has declared intention to become, a citizen of the United States, who has not used his or her homestead right, or who is not then owner of more than 160 acres of land in any one State, can file on any one of the tracts surveyed by the Government. Title to lands cannot be acquired until all payments for water have been made, ten years hence. The law requires a homesteader to see and select his land personally.

There is one warning which should be sounded for the benefit of a certain class of settlers. The man who attempts to make a home on the primeval desert, even with free land and the best irrigation and drainage facilities, requires money to make a successful start. There will doubtless be exceptions to the rule—men who will get work in the locality from the Government or private parties and be able to hold on until their land yields returns, when, by dint of hard work and economical living, they can build their homes, improve their lands, and make their annual payments for water rights. But the average man will need capital in order to bring his farm to a paying stage. This capital he cannot borrow until he gets title to his land, and he cannot get title until he completes payment for his water rights, ten years hence. There is no way in which these payments can be commuted.

It is important that the reader should understand at the outset that a large part of the land to be reclaimed by the national irrigation system is not public domain, open to entry under the land laws, but land in private ownership which the settler may only obtain by purchase from its present proprietors. The proportion of private land varies widely with different projects. There is no locality which is wholly free from it, and there are some localities where nearly all the land is privately owned. The law not only favors, but compels, the subdivision of these private estates into farms of one hundred and sixty acres each. Water rights from government canals cannot be obtained by any one purchaser for a larger area than this. As a matter of fact, the popular farm unit will be much less and there is, of course, no minimum fixed by the law. The irrigated farm may be as small as seller and buyer agree upon.

Eastern readers will, perhaps, desire to be told how it happens that a policy avowedly undertaken with the object of making homes on the public domain should provide for the reclamation of lands in private ownership, and why the first projects undertaken by the Government deal with localities in which the major portion of the land is privately owned.

With few exceptions, the streams on which it is proposed to store water and develop power have long been used for irrigation. Their entire flow at low-water stage is diverted each year into canals already in operation, while an amount of water much in excess of their low-water flow has been claimed and appropriated in accordance with local laws and customs. As a practical question, it is found utterly impossible to store the flood waters of such streams without interfering with vested rights, unless it be frankly conceded at the outset that private lands dependent on this source of supply shall satisfy their reasonable needs from the new works, paying the Government therefor in just the same manner that the settlers on public lands are required to do.

PREPARING HOMES FOR THE PEOPLE

As a rule, there are three classes of lands within reach of every stream in the arid region, *viz.*: first, those owned by earliest appropriators, which have an abundance of water; second, those owned by later appropriators, which have sufficient water for a short time each season and a claim for more water when it happens to be in the stream; third, those who have no water at all, and cannot have until the full storage and pumping possibilities of the stream and locality shall be realized by means of national irrigation. Taking the arid region as a whole, the third class of land is much the largest, and it was for the benefit of this class that the movement was undertaken primarily. But if the Government should shut its eyes to the claims of the second class it would do a grave injustice and have endless litigation on its hands. This would inevitably follow, since the water which the Government proposes to store or to pump is absolutely the only water which can ever be made available for the use of these lands now in private ownership, but receiving only partial and very unsatisfactory irrigation. That is one aspect of the case, and the commonest aspect. But take another where the issue is more sharply defined.

There are streams where every drop of water which can possibly be stored will be required to irrigate lands now in private ownership. These lands were mostly taken up by those who sought to make homes in good faith on the public domain. They settled under laws deliberately enacted by Congress. If those laws proved to be an invitation to disaster, it is certainly not the fault of the homeseekers. They depended for water upon speculative corporations chartered under the law and vested with sweeping franchises and rights in the most precious element of natural wealth. These speculative corporations frequently oversold their supply and more frequently went bankrupt before they had finished their works. Again, the homeseekers were not responsible for the situation in which they found themselves. Indeed, it very often happened that they had paid for their water rights in advance, thus furnishing the speculators with the capital on which to speculate and exploit the unfortunate settlers.

Now, then, to insist that the Government shall appropriate the only water that can ever be brought to these private lands, and take that water away to public lands where nobody lives, would be so palpably unjust that the proposition could not possibly find an advocate or defender among those who know the facts. It therefore becomes necessary not only to irrigate lands of which a part are in private ownership, but it will sometimes be necessary to irrigate lands of which *all* are in private ownership. Not to do so would be an act of injustice, of inhumanity. It would put the Government in the untenable position of punishing one class of its citizens in order that another class may be benefited. With one hand it would hold out the hope of independence to prospective settlers who have not yet left their eastern or foreign homes, while with the other hand it would deprive some of our best and bravest pioneers of their only chance to win the independence they have fought for. This would, of course, be unthinkable.

PREPARING HOMES FOR THE PEOPLE

Salt River Project, Arizona.

The lands to be watered by the great system in this valley, of which the Roosevelt dam in Tonto Basin will be the foundation, are largely in private ownership, but when the present plans of storage and diversion shall be supplemented by the powerful pumping plant which the Government has in contemplation, the total area reclaimed will be almost equally divided between private lands and those now belonging to the public domain.

The Roosevelt dam closes a narrow canyon, which will be flooded to the extent of 14,000 acres. This will create the largest artificial reservoir in the United States. Its capacity will be 400,000,000,000 gallons, or 1,300,000 acre feet,—three times as great as the capacity of the Wachusett reservoir which supplies the city of Boston, and twice as great as the famous Croton dam which supplies the city of New York. Owing to the extreme dryness of the climate, the nature of the soil, and the high degree of evaporative losses, six or seven acre feet per annum are required for irrigation, so that the amount of water stored will increase the usefulness of the Salt River for irrigation to the extent of about 200,000 acres.

The dam is of solid masonry, both foundation and superstructure being built of sandstone. All materials are found on the site, even the cement being manufactured by a special plant located on the ground. The situation, in a deep canyon with lofty, precipitous sides, and with a wide valley above, gives an enormous storage capacity in proportion to the dimensions of the dam. The impounding structure is built in a circular curve, convex up stream, with a total height of 270 feet, a length at the base of only 200, and at the top of 653 feet, with a thickness of 165 feet at the base and 16 feet at the top. Spillways are excavated in solid rock on both sides; these will be bridged, and a road will traverse the bridges and the crest of the dam.

An allotment of $3,600,000 was made for this work from the reclamation fund. The settler must pay about $20 per acre for a water right which represents his share in the ownership of the completed system. This price he is to pay in ten annual instalments without interest. In addition to this, the settler purchasing private lands must pay a price which varies with locality and quality of the soil, ranging from $25 to $100 per acre.*

Colorado River Project.

The Colorado River is nature's great gift to the Southwest and, by the way, it is not in the State of Colorado, but in Arizona, California, and Mexico. For ages it has flowed through the most desolate wastes in North America, a region which has been popularly regarded as worthless and hopeless. During the past thirty years private enterprise has tried to make some use of its turbid flood. Only in the wonderful Imperial Valley of California, and in limited areas below Yuma in Arizona, has there been any result approaching success, and the

* Consult alphabetical index for full description of Salt River Valley and other localities where government reclamation works are building.

SITE OF ROOSEVELT DAM, ARIZONA.

PREPARING HOMES FOR THE PEOPLE

chief value of these experiments has been to demonstrate how little can be done by individual effort and how much could be accomplished if the resources of the Nation were brought to bear upon the problem. The lesson has been learned; the Nation has put its hand to the task.

A great weir, or diverting dam, built after the manner of those which have done good service on the Nile, is now in course of construction in the main channel of the stream, twenty-two miles above Yuma. It will occupy a wide but rocky gorge with solid natural abutments. The height of this weir, which bears the official name of "the Laguna," is but ten feet above low water. Two canals will be taken out at this point, one supplying irrigation to the Arizona side of the river, the other to the California side. The weir will create a settling basin above, in which the heavy silt of the river will be deposited during the greater portion of the year, to be flushed out and sent down to the Gulf at the time of the annual flood. Extensive levees will be built on both banks to project irrigated lands from inundation, and comprehensive drainage systems will be provided wherever necessary as a means of supplementing natural drainage. Ultimately, the floods will be largely stored upon the higher sources of the stream. When the entire water supply is utilized in the most scientific and economical way, nearly two million acres will be irrigated by this great river system, as follows: Above the Grand Canyon (in Wyoming, Colorado, Utah, New Mexico, and Arizona) 470,-000 acres; below the Grand Canyon (in Arizona and California) 811,000 acres; in the Republic of Mexico, 688,000 acres. As navigation must be made subordinate to the higher interests of irrigation, a new treaty will be required with Mexico.

The project, as now planned, will cost about $3,000,-000, but the greater Colorado system which will develop in course of time, and which will comprehend the whole stream from its mountain sources in the north to its outlet in the Gulf, will require an expenditure of, approximately, $20,000,000. Even then it will return its cost several times over every year in dollars and cents, to say nothing of its contribution to civilization in a higher way.

The cost of water rights in the district to be irrigated first will be $35 an acre, which includes the settler's share in the expense of drainage and flood protection. It is likely that the cost will lessen as the project is enlarged to cover greater areas. Much of the choicest land has already passed into private ownership, but a considerable area still remains open to entry under the land laws. The larger portion of the land ultimately to be reclaimed belongs to the public domain. Unimproved land in private ownership will cost the settler from $50 to $100 an acre, in addition to the water right, and is richly worth it. It is susceptible of the most intense cultivation and certain to be closely settled in small farms.

Klamath Project, Oregon and California.

Lying on both sides of the boundary between California and Oregon, the work to be done on "the Klamath project" will be of immense benefit to both those States.

COLORADO RIVER AT HIGH WATER.

PREPARING HOMES FOR THE PEOPLE

This region lies at an elevation of from 4000 to 5000 feet, and is therefore suited only to hardy crops. The arable land is contained in a chain of valleys lying between rocky mountains, all more or less covered with noble pine and hemlock forests. It is a land of snow, and of clear, cold, rushing rivers. The soil consists of decomposed lava and volcanic ash, carried by the streams into the valleys.

The plans involve the construction of two great dams for the creation of reservoirs; the partial drainage of two interstate lakes and the uncovering of the arable marsh lands on their margins; and the use of the abundant water power to pump water from one of the lakes into a near-by valley.

The full development of the work will result in the reclamation of about 300,000 acres of land, at a cost of about $17 an acre. A large proportion of these are public lands. The plans for the project have been approved and funds set aside for the work, which will soon be commenced.

The Uncompahgre Project, Colorado.

In the picturesque and prosperous region on the western slope of the Rocky Mountains, the Government is engaged in completing one of its most notable enterprises, officially known as "the Uncompahgre project." Water is taken from the Gunnison River, conducted through the hitherto impassable mountains by means of a tunnel, and then turned into the channel of the Uncompahgre River, to be diverted into canals and spread over the valleys

THE CONQUEST OF ARID AMERICA

in Montrose and Delta counties. It is an example of bold engineering amply justified by the value of the region to be reclaimed, for there is no fairer spot in Arid America.

The larger portion of the land is in private ownership and commands high prices, owing to the extraordinary returns which are realized in the fruit industry. The project will cost about $2,500,000 and water is expected to be available in 1908. The cost of water rights will be about $25 an acre.

Minidoka and Boise-Payette Projects, Idaho.

Idaho offers ideal conditions for national irrigation enterprise, for it is wonderfully favored in its water supply, as well as in its great valleys of fertile soil. Its comparatively sparse population is another favoring circumstance, since there is the greater opportunity to show large and striking results for the new policy.

The Minidoka project contemplates the reclamation of about 150,000 acres of public land lying on both sides of the Snake River in southern Idaho. This land is near the Oregon Short Line Railroad, in an altitude of about 4200 feet above sea level. A dam is in course of construction across Snake River which will raise the water sufficiently to irrigate the lower valleys. The large water power will be utilized to operate a pumping plant and thus reclaim the higher levels. It is expected that the system will be at the service of settlers in the spring of 1906 and that the cost of water rights will be about $26 an acre.

GOVERNMENT PROJECT IN OREGON.—1. Adams Canal.   2. Tule Lake.

PREPARING HOMES FOR THE PEOPLE

Two townsites have been reserved along the proposed extension of the Oregon Short Line and within the tract to be irrigated. Entries are limited to forty acres within a mile and a half of these townsites, and to eighty acres on all other parts of the tract.

A project of even greater importance is that which is known as the "Boise-Payette," which will irrigate a vast area in the most populous part of the State, near the cities of Boise, Nampa, and Caldwell. The Boise and the Payette are already extensively used for irrigation, but it is proposed to regulate their flow by storage works and largely increase their usefulness. The total area to be irrigated is estimated at 370,000 acres, the greater part of which is in private ownership. It is a case where the Government is combining a number of small systems which had tapped, but not fully developed, the resources of the basin, into one comprehensive and exhaustive system of works. This enterprise will give a tremendous impetus to the growth of southern Idaho and make it one of the greatest irrigated districts in the West.

Milk River, Fort Buford, and Huntley Projects, Montana.

Montana people indulge the confident hope that there will ultimately be more land under government irrigation in that State than in any other part of the Union. Plans already in contemplation will multiply the present agricultural area five or six times over. But if the possibilities are great, the day of realization is placed some distance in the future because of international

THE CONQUEST OF ARID AMERICA

difficulties, as well as by the magnitude of the opportunity, which calls for a very large expenditure.

The most important project yet outlined in Montana is that which proposes to utilize the waters of Milk River near the Canadian border. As in the case of the Rio Grande in New Mexico and the Rio Colorado in California, international questions have arisen here which delay development. Thus far the Government has undertaken to irrigate only so much land as may be done from the usual flow of the stream without storage. This amounts to 60,000 acres. It is mostly public land and can be reclaimed at a cost of about $15 an acre, which makes it one of the cheapest propositions open to the settler. The district is traversed by the Great Northern Railway for a distance of seventy-five miles, the chief seat of activity being near the town of Malta. When the larger development is undertaken, water will be stored in St. Mary Lakes, whence it will be conducted by a large canal into the Marias River, a tributary of the Milk. The completion of the whole comprehensive project will involve a cost of between $3,000,000 and $4,000,000.

The Fort Buford project, although generally credited to North Dakota, lies chiefly in eastern Montana. It will be watered from the Yellowstone. Of the total area (64,144 acres) 27,404 acres are in private ownership, 22,665 in public ownership, and 14,075 owned by the Railroad. The lands lie in a long strip on the west side of the Yellowstone, sixty-five miles long and five miles wide at the widest part. The cost of water rights will be about $30 an acre.

PREPARING HOMES FOR THE PEOPLE

Construction has begun on the reclamation of 35,000 acres with the waters of the Yellowstone and Bighorn. This is known as "the Huntley project." The Indians are to be provided with land in severalty, after which there will be an opportunity for settlers. Water rights will cost about $30 an acre, a cost amply justified by the value of the lands.

North Platte Projects, Wyoming and Nebraska.

What is known as the "North Platte" or "Interstate project" aims at the reclamation of a vast area lying on both sides of the North Platte River for a distance of two hundred and ten miles. There are over half a million acres in Wyoming and about a quarter of a million acres in Nebraska to be prepared for cultivation by this single enterprise. The scheme involves the construction of three permanent dams, two temporary diversion dams, three outlet tunnels in solid rock formation, and several great distributing canals. The first and greatest of these dams is "the Pathfinder," so named in honor of General Fremont, situated in the great canyon of the Platte, ten miles above Alcova, Wyoming. It will require four years to complete it.

Truckee-Carson Project, Nevada.

What is known as "the Truckee-Carson project" will ultimately irrigate 375,000 acres at a cost of about $9,000,000. Nine years will be required to bring it to completion. The portion of the works put into operation on

June 17th, 1905, will distribute water to about 50,000 acres and represents a cost of about $1,750,000.

The main canal now in operation diverts the water from the channel of the Truckee at a point twenty-four miles east of Reno and conveys it through the divide to the Carson River, a distance of thirty-one miles. This canal has a capacity for the first six miles of its course of 1400 cubic feet per second, or 70,000 miner's inches under a four-inch pressure, and, for the remainder of its course, of 1200 cubic feet per second. There are three tunnels, all lined with concrete, as are two miles of the canal outside of the tunnels. The main canal discharges its water into a natural reservoir on the Carson and flows thence four and one-half miles to the diversion dam at the head of the distributing system, where it is led out upon the land in two wide-reaching canals, one on each side of the river. The canal on the north side has a capacity of 450 cubic feet per second; that on the south side, a capacity of 1500 cubic feet per second. With their main branches, these waterways will ultimately have a total length of over 90 miles, while the laterals and drain-ditches to be constructed in Carson Sink Valley alone will aggregate fully 1200 miles.

The dam in the Carson at the head of the distributing system is something to bring a smile of satisfaction to the faces of those who have known the crude brush dams of the pioneers and the endless difficulties which arose from them. This government dam is a solid concrete structure, built for a thousand years. It furnishes an absolute guaranty of a permanent water supply to

ONE OF UNCLE SAM'S TUNNELS, NEVADA.

PREPARING HOMES FOR THE PEOPLE

the settlers. This, indeed, is the character of all the work the Government has done.

The land to be irrigated is located in a number of valleys along the Truckee and Carson Rivers, extending on each side from the Central Pacific Railroad, the greatest distance from the road being twenty-five miles. The soil is adapted to alfalfa and other forage crops, potatoes, onions, beets, and other vegetables, apples, pears, berries, and similar hardy fruit.

Nearly all the land now irrigated was public property until recently filed upon, after the works were undertaken. Some of it is still open, but this condition will not continue long. No price is charged for the land, except filing fees, which are nominal. But the settler must repay the cost of irrigation in ten annual instalments, without interest. This amounts to $26 an acre, of which about $10 an acre has been incurred by the provision of drainage facilities, made imperatively necessary as a means of removing the heavy alkali deposits. The settler is fortunate to be able to make his home where conditions have been scientifically ascertained in advance and where the best engineering skill, together with abundant capital, have been available to make the most thorough preparation for his success.

Hondo Project, New Mexico.

The construction of efficient and enduring reclamation works in Arizona and New Mexico presents problems peculiar to those regions. The torrential character of the streams, with their powerful floods in the rainy season, the heavy silt carried, the absence of rain during the major portion of the year, and the consequent dependence of agriculture upon water storage, make special constructions necessary in all branches of the work.

The principal project upon which the Government is engaged in New Mexico, is known as "the Hondo project," and is situated on the Hondo River, a tributary of the Pecos, twelve miles southwesterly from Roswell, in Chaves County. Roswell is the seat of this county, and is on the Pecos Valley & Northeastern Railroad. The tract to be reclaimed consists of about 10,000 acres, almost entirely in private ownership. The total cost is estimated at $275,000, or $27.50 per acre.

The works consist of a storage reservoir on the north side of the Hondo River, with appropriate canals. The Rio Hondo Reservoir Water Users Association has been formed, has secured the subscription of 10,000 acres, and is co-operating with the government engineers. The owners of private irrigation schemes opposed the government project strongly, and the opposition ceased only when a flood washed out the private dams and canals; then the settlers begged the Government to come in and save their apple and peach orchards. This the Reclamation Service is doing, and has hastened the work as much as possible. The work is now well forward, and the water will soon be ready for delivery from the new canals.

Bismarck and Buford-Trenton Projects, North Dakota.

In western North Dakota a number of projects are being developed for the reclamation of flat lands lying

DETAILED VIEW OF TOP OF WASTE-WAY GATES, TRUCKEE-CARSON PROJECT, NEVADA.

at considerable elevations above the Missouri River. These projects involve the most extensive use of pumping plants yet undertaken, but this obstacle will be overcome by the use of the cheap and abundant coal taken from public lands near by.

The Bismarck project, near the city of Bismarck, involves the reclamation of about 15,000 acres, mostly private lands, $250,000 having been set apart for the work.

The Buford-Trenton pumping project will reclaim lands on the north side of the Missouri River, between the State boundary and Williston. About 18,000 acres are reclaimable, at a cost of a little over $16 per acre.

Malheur Project, Oregon.

No State offers greater attractions for national enterprise, in the way of large areas of fertile desert lands and abundant water supply, than Oregon. It is, perhaps, unfortunate for the State that so many of the best irrigation propositions were allowed to be undertaken by private enterprise under the "Carey Act," thus shutting the Reclamation Service out of a large part of its best field of operations.

The enterprise known as "the Malheur project" is the principal government undertaking in Oregon. The lands lie in the east-central part of the State, on the Malheur River, in the valley of the same name, on both sides of the river, and extending from the town of Vale to the Snake River. About 100,000 acres will be reclaimed, at a cost of about $30 an acre.

THE CONQUEST OF ARID AMERICA

The elevation is from 2150 to 2350 feet. The lands are said to be exceptionally fertile, and lie partly in the valley and partly on the benches. Nearly half are public lands. The district to be irrigated lies near Payette, Idaho, on the opposite side of the river, and the conditions are similar.

The method of reclamation will be by storing the summer flow of the Malheur and its tributaries in reservoirs at the head of the drainage basin, which can be done without injury to present water users. The present plans contemplate works on the Owyhee and Willow Creek, both being tributaries of the Malheur, and are capable of considerable extension.

Bellefourche Project, South Dakota.

Lying on both sides of the Bellefourche River, in Butte and Meade counties, South Dakota, lies a tract of irrigable land consisting of about 85,000 acres. The work of its reclamation is being undertaken, under the name of the "Bellefourche project." This land is yet principally public land, and has had little value hitherto for anything except grazing. The cost of its reclamation will be about $30 an acre. The region is well supplied with railroad facilities, three lines traversing or passing near it.

A large basin east of the town of Bellefourche is to be converted into a reservoir by the construction of an earth embankment, riprapped with rock, across Owl Creek, a tributary of the Bellefourche. Other water supplies will be obtained in a similar manner from near-by creeks and brought in by feeder canals.

PREPARING HOMES FOR THE PEOPLE

Palouse River Project, Washington.

"The Palouse River project" is the principal project now under way in Washington. The lands lie between the Columbia River on the west and the Snake River on the south, and extend from the town of Connell to the town of Pasco. About 100,000 acres are to be reclaimed, at a cost of about $30 an acre. Most of the lands are in private ownership.

The summer flow of the Palouse River is insufficient to irrigate these lands, and it is therefore necessary to store the flood waters. This will be done by a series of reservoirs. The lands to be reclaimed are of variable character, those in the upper part of the tract having a light soil, about the consistency of flour, such as is so prevalent in arid regions. In the lower portion, this is mixed with a large proportion of sand.

At Connell and Eltopia the conditions are favorable for pumping water to higher elevations with power developed by a drop in the main canal. The region is surrounded by excellent lands on which dry wheat farming is practiced with varying success. These lands, unfortunately, lie too high to be irrigated from this source. The transportation facilities are good.

Shoshone Project, Wyoming.

After years of effort to bring about the reclamation of his beautiful domain by means of private enterprise, Col. William F. Cody has had the happiness of seeing Uncle Sam take hold of the work. This is what is officially known as "the Shoshone project." It involves the reclamation of about 160,000 acres of fertile lands in Bighorn County, on the north side of the Shoshone River, near the town of Cody.

Nearly all this land had been segregated under the "Carey Act" when the government engineers began to consider the project; but the State had not yet begun the work, and therefore relinquished its rights. Col. Cody also transferred his rights in the waters of the Shoshone River to the Government.

The main diversion dam is at the head of the Box Canyon, a gorge 210 feet deep, 85 feet wide on the bottom and 165 on top. The water will be spread over the bench lands lying on the north side of the river for 45 miles. These works are among the largest and most important yet undertaken by the Reclamation Service.

Where More Homes Will be Made.

The foregoing are all the projects which have been formally approved by the Secretary of the Interior and money set aside from the reclamation fund for their construction, or work already begun.* There are secondary projects in all the arid States and Territories, in various conditions of forwardness, many of which involve large engineering constructions, and a few of which cover vast areas. It is impossible to give them more than the briefest reference here.

Arizona has the San Carlos project, which was carefully considered, but laid aside for the Salt River project. It will reclaim 100,000 acres on the San Carlos

* Up to September, 1905.

PREPARING HOMES FOR THE PEOPLE

River. The basin of the Little Colorado River, and other promising localities, are also being studied.

It is possible that the largest of all the government irrigation projects may grow out of the investigations now being made in the Sacramento Valley, in California. It would be difficult to convey an adequate idea of the immensity of the task.

Investigations begun on the western slope of San Diego County are the first undertaken under the new plan of co-operation between local irrigation districts and the Reclamation Service. If they bring results, this may be the beginning of a new day in the irrigation development of Arid America.

Colorado has a number of promising projects. On the Grand River, near the City of Grand Junction, 60,000 acres are to be reclaimed. The White River project involves 90,000 acres in Rio Blanco and Routt counties, near the town of Meeker. A chain of reservoir sites are being planned in the Rocky Mountain region of southwestern Colorado, for the purpose of controlling the flow of the Colorado for the benefit of irrigation upon its lower course.

The coming project in Idaho is known as "the Dubois project," for the reclamation of about 200,000 acres in Blaine and Fremont counties from the headwaters of the Snake River.

Kansas has no projects now under construction, but the Government is investigating the underground flow of the Arkansas River in the vicinity of Garden City and an allotment of $49,903 has been made for the purpose of installing a pumping plant by which it is hoped to reclaim about 2,000 acres. The success of this undertaking is expected to be very useful in the teaching of scientific methods to the farmers.

Montana has a number of promising projects. That on the Sun River is expected to reclaim about 200,000 acres of public lands.

Reconnaissances are being made of the Niobrara and Snake Rivers, northeastern Nebraska.

Extensive surveys of the Walker and Humboldt rivers are being made in Nevada, as well as a general investigation of the problems of seepage and drainage in that State.

New Mexico has a number of promising projects. The Urton Lake project, in Guadalupe and Chaves Counties, will open about 60,000 acres of public land to settlement. The La Plata project will reclaim 50,000 acres in San Juan County. The Las Vegas grant, near the City of Las Vegas, is to be surrendered to the Government by its owners in return for its irrigation, and it will then be open to settlers. This is the first instance of the kind in the history of the irrigation work.

Work in North Dakota consists of a general search for reservoir sites and feasible projects, gaging of streams, and examination of pumping projects in the western portion of the State. The large areas of good coal which is easily to be mined upon public lands, makes this region peculiarly adapted to the establishment of successful pumping plants.

Oklahoma is awake to the advantages of national irrigation. Investigations are being made of the watersheds of the Cimarron, Canadian, and other rivers, and for reservoir sites and underground waters in the arid region of its western panhandle. The "Otter Creek project" is expected to reclaim about 40,000 acres of Valley land lying south of Mountain Park.

The secondary projects in Oregon are principally for the reclamation of high-lying valleys in more or less remote districts. Some of them have had to be abandoned on account of prohibitive cost of the work. A very large area is involved. Those projects which seem to be yet alive at this time are the Silver Lake project near the lake of that name in Lake County; the Ana River project, between Summer and Silver lakes, in the same county; the Chewaucan project, near Paisley, Lake County; and the Silver Creek project, near the town of Riley, in Harney County.

South Dakota has the Cheyenne project, by which it is expected that 40,000 acres will ultimately be reclaimed, near the town of Edgemont.

It is matter of regret that the Utah projects cannot be described in detail. The whole water system of the State is being studied and works planned upon a comprehensive scale. The first work undertaken will doubtless be the lowering of the level of Utah Lake, to diminish the evaporation and increase its sources of supply. It is estimated that by this means the irrigated area in the Jordan Valley can be increased 60,000 acres. Other projects are known as the "Bear Lake" and "Strawberry Valley" projects, both extensive and important.

Two coming projects sustain the hope of the people of Washington. The Okanogan project contemplates the reclamation of a large tract in the county and from

THE CONQUEST OF ARID AMERICA

the river of that name. The Big Bend project is one of the largest in the Arid West, and may ultimately involve a million acres, in the bend surrounded by the Columbia River on the west, the Snake River on the south, and the high wheat lands of the Palouse region on the north and east. The Yakima Valley, Priest Rapids, and Kootenai Basin, are regions where the water supply and lands are being investigated.

In Wyoming, the Lake De Smet project, in Johnson and Sheridan counties, is under consideration. Large areas are involved, principally in private ownership. In the Big Bend of the Wind River Mountains, a part of the Shoshone Indian Reservation, there are about 230,000 acres of good lands, the reclamation of which may be undertaken soon.

The plans of the Reclamation Service are developing so rapidly that any record of its work must be regarded as merely temporary. Its investigations are being prosecuted constantly and it may very likely happen that some of its most important undertakings in the early future will be such as are not mentioned in these pages. Enough has been said, however, to indicate the varied character of its work and to furnish the intending settler with information of some value.

CONCLUSION

MAN'S PARTNERSHIP WITH GOD

SOME one has said that God never made a world; that He started several, including the one on which we dwell, but that He depends on man, working in partnership with Him and in harmony with the laws of the universe, to bring the world to completion.

There are conditions in Arid America which make men peculiarly conscious of their partnership with God or Universal Purpose. They seem, indeed, to begin where God left off and to go forward with the actual material creation of the world. Here, the English-speaking race entered into a new environment. Nature had done what it would, then withdrawn and left its unfinished task to the ingenuity of man.

The waste of desert and mountain has been unsympathetically called "the land that God forgot." Time will show, and already time has begun to show, that above all other sections Arid America is the God-remembered land. He evidently remembered that somewhere there must be a place where man should become supremely alive to his divinity—that somewhere he must be driven by the club of necessity into a brotherhood of labor—that somewhere the material must be blended with the

MAN'S PARTNERSHIP WITH GOD

spiritual until man should stand erect, the conscious partnership of the universe.

The first thought that comes to the man of insight on viewing the region is the utter futility of individual effort in the stupendous struggle with nature. There is soil fertile and enduring beyond that of any other land, but nature neglects to water it with unfailing rains. The men of an earlier and more superstitious age would have fallen on their knees in prayer; but an eternity of such prayers would bring no response from the smiling sky. Men learn from their environment a better way to pray.

Conforming their methods to the laws of the universe and entering into glad partnership with God, they follow the torrential stream to its mountain sources, discover the reservoir sites which nature provided at the right elevation to command the valley and to furnish power with which to bring the hidden water from the bowels of the earth; and thus, blending science with religion, and the material with the spiritual, their prayers are answered with fullest measure of blessings,—blessings, as we saw in earlier pages, infinitely superior to those which come to other lands by dependence upon rainfall.

This process not only brings men close to Divinity in nature; it brings them close to Divinity in man. Brotherhood becomes compulsory,—they must work together, must work with and for each other rather than against each other. They are enlisted in a common effort for the accomplishment of a common good. The welfare of each is the concern of all. "Bear ye one another's burdens" is the stern mandate written on the face of nature

THE CONQUEST OF ARID AMERICA

and borne on the voice of the flood. Men are compelled by force of circumstances to labor *en masse* and in harmony with the universe.

But it is not only when men are working together at a common task that they are conscious of the divine partnership. The man who works intelligently in creating his irrigated farm with the raw materials of land and water, knows that in this smaller sphere he is engaged in finishing the world. He feels himself to be an instrument in the process of evolution. His constant study is to learn the secret of his environment and to shape his industry in such a way as to attain the best results. If he were tilling his ancestral acres in Maine or Kentucky, he would follow the methods of his forefathers, but would not be driven by necessity to experiment with the amount of moisture required for different crops, the manner of applying it to orchard, garden, and field, the temperatures of water and soil, or the drainage of his land by natural and artificial means. In other words, he would not be conscious of his part in the practical evolution of nature's raw materials into civilization's finished product.

There have been countless instances of men who came from the stagnant life of established communities to settle in such progressive colonies as Greeley, Colorado, or Riverside, California, and developed into scientists of the most practical sort. They had no choice in the matter; they must adapt themselves to new conditions in order to prosper. In so doing, they necessarily co-operated with Universal Purpose in finishing the world.

The same influence is dominant in the formation of institutions. Laws and customs must conform to en-

MAN'S PARTNERSHIP WITH GOD

vironment, and to the work to be done under the conditions which environment imposes. The first requirement is scientific knowledge of these conditions; the next, adaptation of institutions in conformity with these ascertained facts.

To proceed in the making of your farm, in the development of a great region, in the formation of institutions, by knowledge rather than by chance, is a profoundly religious thing. Irrigation, for example, is a religious rite. Such a prayer for rain is intelligent, scientific, worthy of man's divinity. And it is answered. To put knowledge in place of superstition is the first step which men take in entering into partnership with God.

All labor done in the spirit of this partnership is a blending of the material with the spiritual. It is inspired by knowledge of universal law; it aims at the accomplishment of results in line with Universal Purpose. Then labor ceases to be drudgery and becomes beautiful.

And the relations of brotherhood which men necessarily sustain to each other under such conditions represent the essence of religion, for they are inspired by the love of humanity.

Men have too often proceeded in defiance of universal laws, so far as they could do so and exist. They have wasted the bounty of nature where they should have conserved it; defaced the landscape where they should have beautified it; dissipated their strength in fighting each other where they should have combined their strength and worked together; degraded toil where they should have ennobled and glorified it. All this they have done because

THE CONQUEST OF ARID AMERICA

no imperative necessity compelled them to make the acquaintance of God and work hand-in-hand with Him in finishing the world.

It is the fortune of Arid America to be so palpably crude material that it can not be used at all, save upon the divine terms.

THE END.

APPENDIX I

NOTE AS TO METHODS OF IRRIGATION

To those who are unfamiliar with the life of the arid region the actual process of irrigation seems a deep mystery. They regard it as an effort to overturn the laws of nature, and think it must be accompanied by a struggle as severe as it is inscrutable. But irrigation is, after all, a perfectly natural, and even a familiar, process. The man who waters his plat of grass and the woman who waters her door-yard pansies are irrigators in a rude and humble way. The citizen who grumbles at the sight of withered lawns in a public park during a dry summer yearns for irrigation without knowing it. A generation which has harnessed the lightning should see nothing incongruous in the ancient expedient of storing the rain and distributing it to meet the varying needs of plants which nourish human life.

The control of water for irrigation in the West presents about the same problems to the engineer as the control of water for domestic purposes in large cities and towns. The water must be diverted from a flowing stream at a level sufficiently high to command the territory to be irrigated; or it must be impounded in reservoirs at a season of floods or unusual flow, such as occurs everywhere when the ice and snow are melting; or it must be sought in the bowels of the earth by means of wells and lifted to the surface by pumps, except

APPENDIX I

in the case of artesian waters, which flow out of the mouth of the well by reason of their own pressure.

The principal difference between securing a supply for domestic and for agricultural purposes is that in the case of the former the water must be as pure as possible, while in the case of the latter the impurities which gather in ponds and streams have a distinct commercial value as fertilizers. The sewage of Paris is used for irrigation purposes with wonderful effect. The same thing is done at Los Angeles, and doubtless will be done in many places hereafter. Neither is it necessary, as a rule, to make such elaborate provision for the distribution of water through underground pipes in the case of agriculture as in that of domestic water supply. In the vast majority of instances irrigation water flows in open channels. Where it is otherwise it is because the precious fluid is scarce, and therefore dear, so that every drop must be guarded against loss by evaporation or by seepage into the ground.

Irrigation works in the West range from rude and simple ditches, taking their supplies from mountain brooks where the water has been diverted by means of small brush dams, to great masonry walls which block the outlet of deep canyons, holding back the water, which is transported through pipes, flumes, and cemented ditches to rich lands miles away. In the one case the works have been constructed by a small association of farmers, using their own labor and teams; in the other, millions of eastern and foreign capital have been invested. In both cases the water is led through main canals to central points in the territory to be reclaimed. These mains are of all sizes, depending entirely upon the volume of water required. They are frequently not more than six feet wide, though some of the canals in the San Joaquin Valley are one hundred and twenty feet in width. From these mains lateral ditches reach out in various directions. The individual farmer taps the lateral with a shallow ditch, usually made

APPENDIX I

with a plough, and thus conducts the water where he wants it through his own private system of distributers. The management of the waters, when the system has once been perfected, is so simple that a child can attend to it. Furnishing arid lands with irrigation facilities is really a less formidable task than supplying cities with water for domestic and fire purposes. The one process is no more mysterious and unnatural than the other.

Although irrigation is both ancient and universal, the Anglo-Saxon never dealt with it in a large way until the last half-century, when he found it to be the indispensable condition of settlement in large portions of western America, Australia, and South Africa. Through all the centuries of the past the art has been the exclusive possession of Indian, Latin, and Mongolian races. Its earliest modern traces in this country are found in the small gardens of the Mission fathers of southern California. They brought the method from Mexico and taught it to the Indians. But the real cradle of American irrigation as a practical industry is Utah.

In the hands of the Indians and Mexicans of the Southwest irrigation was a stagnant art, but the white population studied it with the same enthusiasm it bestowed upon electricity and new mining processes. The lower races merely knew that if crops were expected to grow on dry land, they must be artificially watered. They proceeded to pour on the water by the rudest method. The Anglo-Saxon demanded to know why crops required water, and how and when it could best be supplied to meet their diverse needs. He has thus approached by gradual steps true scientific methods, which are producing results unknown before in any part of the world.

The earliest method of irrigation is known as "flooding," and is generally applied by means of shallow basins. A plot of ground near the river or ditch from which water is to be

APPENDIX I

drawn is inclosed by low embankments called checks. These checks are multiplied until the whole field is covered. The water is then drawn into the highest basin, permitted to stand until the ground is thoroughly soaked, and then drawn off by a small gate into the next basin. This process is repeated until the entire field is irrigated. This is the system practised on the Nile, where the basins sometimes cover several square miles each, while in the West they are often no more than four hundred feet square. There is both a crude and a skilful way to accomplish the operation of flooding, and there is a wide difference in the results obtained by the two methods. The Indian and Mexican irrigators, in their ignorance and laziness, seldom attempt to grade the surface of the ground. They permit water to remain in stagnant pools where there are depressions, while high places stand out as dusty islands for generations. All except very sandy soils bake in the hot sunshine after being flooded, and the crude way to remedy the matter is to turn on more water. Water in excess is an injury, and both the soil and the crops resent this method of treatment.

The skilful irrigator grades the soil to an even slope of about one inch to every hundred inches, filling depressions and levelling high places. He "rushes" the water over the plot as rapidly as possible, and when the ground has dried sufficiently cultivates the soil thoroughly, thus allowing the air to penetrate it. The best irrigators have abandoned the check system altogether, and have invented better methods of flooding the crops. Cereals and grasses must always be irrigated by flooding, but the check system seems likely to remain only in the land of Spanish speech and tradition, where it was born. In Colorado wheat and grass are generally irrigated by a system of shallow plough furrows run diagonally across a field. The water is turned from these upon the ground, and permitted to spread out into a hundred small rills, following the contour of the land. Some farmers be-

APPENDIX I

stow great pains upon this method, and succeed in wetting the ground very evenly. Another method of flooding fields is now much used in connection with alfalfa, a wonderful forage plant extensively cultivated throughout the arid region. This produces three crops a year in the north and six crops in the south, and is not only eaten by stock, but by poultry and swine. To find the best method of watering this valuable crop has been the object of careful study and experiment in the West. It is now accomplished by means of shallow indentations or creases, which are not as large as furrows, but accomplish the same purpose. These are made by a simple implement at intervals of about twelve inches. They effect a very thorough and even wetting of the ground.

The scientific side of irrigation is to be studied rather in connection with the culture of fruit and vegetables than with field crops. It is here that the English-speaking irrigators of the West have produced their best results. California has accomplished more than any other locality, but nothing was learned even there until the man from the North had supplanted the Spanish irrigator. The ideal climatic conditions of California attracted both wealth and intelligence into its irrigation industry. Scarcity of water and high land values operated to promote the study of ideal methods. Where water is abundant it is carried in open ditches, and little thought is given to the items of seepage through the soil and loss by evaporation. Under such conditions water is lavishly used, frequently to the injury rather than the benefit of crops. But in southern California water is as gold, and is sought for in mountain tunnels and in the beds of streams. A thing so dearly obtained is not to be carelessly wasted before it reaches the place of use. Hence, steep and narrow ditches cemented on the bottom, or steel pipes and wooden flumes, are employed.

This precious water is applied to the soil by means of small furrows run between the trees or rows of vegetables.

APPENDIX I

The ground has first been evenly graded on the face of each slope. The aim of the skilful irrigator is to allow the water to saturate the ground evenly in each direction, so as to reach the roots of the tree or plant. The stream is small, and creeps slowly down the furrow to the end of the orchard, where any surplus is absorbed by a strip of alfalfa, which acts like a sponge. The land is kept thoroughly cultivated, and in the best orchards no weed or spear of grass is ever seen; the water is too costly to waste in the nourishment of weeds. Moreover, it is desired to leave the soil open to the action of air and sunshine. Nowhere in the world is so much care given to the aëration of the soil as in the irrigated orchards and gardens of the West. Too much water reduces the temperature of the soil, sometimes develops hard-pan, and more frequently brings alkali to the surface. For these reasons modern science has enforced the economical use of water, reversing the crude Mexican custom of prodigal wastefulness. The success of the furrow method depends somewhat upon the texture of the soil, and there are places where it cannot be used at all. Such localities are not considered favorable to fruit culture.

Of late years in California the application of water by furrows has been brought to a marvellous degree of perfection. What is known as the "Redlands system" is the best type of irrigation methods known in the world. Under this system a small wooden flume or box is placed at the head of the orchard. An opening is made opposite each furrow, and through this the water flows in the desired quantity, being operated by a small gate or slide. The aperture regulates the flow of water accurately, and the system is so simple that, after it is once adjusted, its operation is as easy as the turning of a faucet. The farmer who grows his crops on a fertile soil, under almost cloudless skies, with a system controlling the moisture as effective as this, may be said to have mastered the forces of nature. The quality of the fruit has

APPENDIX I

improved immensely since the California methods were perfected. Every fruit-grower realizes that the profit in his business comes mostly from the first grade of fruit. Scientific irrigation makes it possible for him largely to increase the percentage of the best fruit, and the difference which this makes in the earning capacity of his acres is surprising.

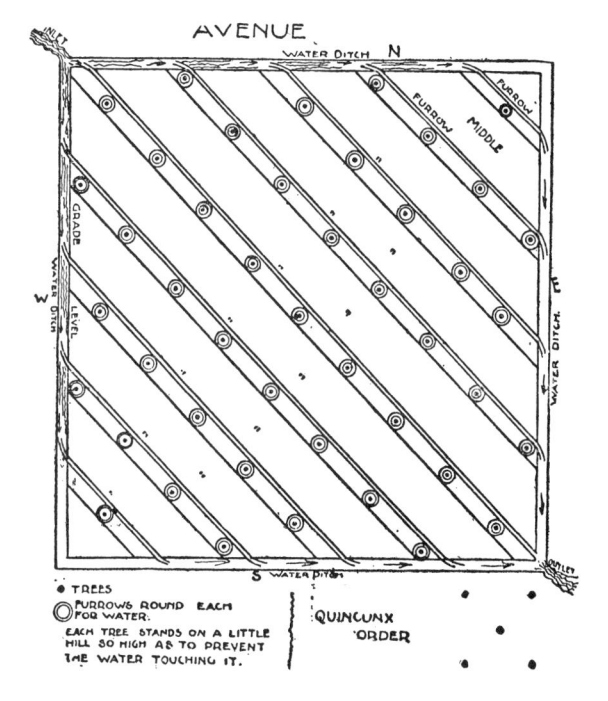

Other methods of furrow irrigation have been devised which are scarcely less perfect than those used in the California orange districts. One of the best of these is the result of the labors and experiments of Professor A. E. Blount, of the Agricultural College at Las Cruces, New Mexico, and is illustrated in the accompanying diagram. In this case the water is carried in small open ditches, and the furrows are

APPENDIX I

extended in circles around each tree, but the water is never allowed to touch the bark. This method is, perhaps, better adapted to the general needs of the arid region than the more expensive plan of the Californians. It is interesting to note that the modern New Mexico method was developed in the midst of Indian and Spanish settlements, which still pursue the methods of antiquity without the slightest abatement of their evils.

One of the most interesting results of irrigation, in a social and economic way, is its influence upon the density of population. The densest population in the eastern States obtains in Rhode Island, where there are two hundred and seventy-six persons to each square mile. In a representative locality of southern California, which is in the midst of the older settled irrigated districts, there are five hundred persons to the square mile, practically all of them engaged in horticulture by means of irrigation. The Nile lands of Egypt support a population of twelve hundred and twenty-seven persons to the square mile. There is, therefore, no risk whatever in predicting that the arid lands of the West will ultimately sustain much the densest population in the United States.

While the perfect conditions for the irrigation industry exist only in an arid land, there is no doubt that the same methods can and will be used largely in the eastern portion of the United States. There is seldom a year when large districts east of the Mississippi do not suffer heavy losses from the lack of rain at the time when it is needed. What irrigation can accomplish under such conditions has been strikingly illustrated by Dr. Clarke Gapen, Superintendent of the State Insane Asylum at Kankakee, Illinois. This gentleman became convinced that if he could control the moisture during the dry period of the Illinois summer, he could readily produce, on the farm operated in connection with the public institution, the large quantities of late vegetables which

APPENDIX I

he had been in the habit of purchasing for cash. He obtained an inexpensive pumping-plant and engaged the services of a practised irrigator. The result was the saving of an annual expenditure of fifteen thousand dollars for farm products, so that the irrigation system more than paid for itself the first year. Dr. Gapen has stated that the experiment convinced him "that if land is worth one hundred dollars per acre in Illinois without irrigation, it is worth five hundred dollars with it." If this enterprising official had suggested ten years before that irrigation was necessary in Illinois, he would have been regarded as a proper subject for one of the padded cells in his own asylum.

The local application of irrigation is now frequently discussed in the farm journals of Ohio, New York, and other eastern States. The art has been employed for a number of years in the most profitable market-gardens about Boston. The western friends of irrigation have the utmost confidence that during the next century their methods will be extensively adopted in the East, resulting in a very great reduction of the average farm unit, in the assurance of much larger and better crops, and in wonderful social gains.

APPENDIX II

THE NEWLANDS BILL, AND THE ACT OF JUNE 17, 1902

THE following is the text of the original Newlands Irrigation Bill, introduced January 26, 1901, before the appointment of the famous Committee of Seventeen and nearly eight months prior to the accession of President Roosevelt:

"*Be it enacted by the Senate and House of Representatives of the United States of America in Congress assembled*, That all moneys received from the sale and disposal of public lands in the arid and semi-arid States and Territories beginning with the fiscal year ending June thirtieth, nineteen hundred and one, excepting those set aside by law for educational purposes, shall be reserved and set aside for the creation of a fund in the Treasury, to be known as the 'arid land reclamation fund,' for the construction of reservoirs and other hydraulic works for the storage and diversion of water for the irrigation and reclamation of arid land.

"Sec. 2. That the Secretary of the Interior, by means of the Director of the Geological Survey, be, and hereby is, directed to continue the examination of that portion of the arid region of the United States where agriculture is carried on by means of irrigation as to the advantages for the storage of water for irrigating purposes, of the practicability of constructing reservoirs, together with the capacity of the streams and the cost of construction and capacity of reservoirs, and such other facts as bear on the question of storage of water for irrigating purposes as required by the Act approved March twentieth, eighteen hundred and eighty-eight and also to investigate the

APPENDIX II

practicability of diverting large rivers by means of tunnels or other works, and of providing supplies by means of artesian wells.

" Sec. 3. That the Director of the Geological Survey shall from time to time make reports to the Secretary of the Interior as to each of various proposed reservoirs, diverting canals, or other methods of procuring water, said reports to show the location, cost of construction, quantity, and location of such land as can be irrigated, as well as the other facts relative to the practicability of the enterprise.

" Sec. 4. That upon the filing of such report the Secretary of the Interior may, in his discretion, withdraw from public entry the lands required for reservoir or other hydraulic works, together with the public lands which it is proposed to irrigate from such works.

" Sec. 5. That upon the determination by the Secretary of the Interior that each of the said projects of reclamation is practicable he shall cause to be let, upon proper public notice, contracts for the construction of the same, in whole or in part, payments to be made from the arid land reclamation fund : *Provided*, That no such contract shall be let until the necessary funds are available: *And provided further*, That in all construction work eight hours shall constitute a day's work and none but citizen labor shall be employed.

"Sec. 6. That upon the completion of each irrigation project, the total cost thereof shall be ascertained and the amount divided pro rata per acre of the lands to be irrigated thereby, and that said amount shall be made a charge against the lands as the cost of a right to the use of water from said system of irrigation, and that said public lands shall be subject to homestead entry, after notice by the Secretary of the Interior, upon the condition that in addition to the requirements of the homestead Act the entryman shall make payment to the Government of the cost per acre of water right as above ascertained, said payment to be made in not to exceed ten annual instalments, and each entryman shall be limited to the entry and settlement of eighty acres, or such lesser amount as the Secretary of the Interior may designate, and the moneys thus re-

APPENDIX II

ceived shall be covered into the arid land reclamation fund: *Provided further,* That the right to the use of water shall be perpetually appurtenant to the land irrigated, and beneficial use shall be the basis, the measure, and the limit of the right.

"Sec. 7. That in case the water thus provided shall be more than sufficient for the reclamation of the public lands, or if land in private ownership has been found by the survey above authorized to be better suited for the utilization of the stored or divided waters, or if there is a sufficiency for both, then the right to use such water may be sold at the rate as above ascertained and under the same terms; but no water right shall be granted to any landowner or occupant for an amount exceeding eighty acres. The proceeds of such sales shall be covered into the arid land reclamation fund.

"Sec. 8. That the following shall be considered as arid land and semi-arid land States and Territories within the meaning of this Act: Arizona, California, Colorado, Idaho, Kansas, Montana, Nebraska, Nevada, New Mexico, North Dakota, Oklahoma, Oregon, South Dakota, Utah, Washington, Wyoming."

Following is the full text of the present law:

"*Be it enacted by the Senate and House of Representatives of the United States of America in Congress assembled,* That all moneys received from the sale and disposal of public lands in Arizona, California, Colorado, Idaho, Kansas, Montana, Nebraska, Nevada, New Mexico, North Dakota, Oklahoma, Oregon, South Dakota, Utah, Washington, and Wyoming, beginning with the fiscal year ending June thirtieth, nineteen hundred and one, including the surplus of fees and commissions in excess of allowances to registers and receivers, and excepting the five per centum of the proceeds of the sales of public lands in the above States set aside by law for educational and other purposes, shall be, and the same are hereby, reserved, set aside, and appropriated as a special fund in the Treasury to be known as the 'reclamation fund,' to be used in the examination and survey for and the construction and maintenance of irrigation works for the storage, diversion, and development of waters for

APPENDIX II

the reclamation of arid and semi-arid lands in the said States and Territories, and for the payment of all other expenditures provided for in this Act: *Provided,* That in case the receipts from the sale and disposal of public lands other than those realized from the sale and disposal of lands referred to in this section are insufficient to meet the requirements for the support of agricultural colleges in the several States and Territories, under the Act of August thirtieth, eighteen hundred and ninety, entitled ' An act to apply a portion of the proceeds of the public lands to the more complete endowment and support of the colleges for the benefit of agriculture and the mechanic arts, established under the provisions of an Act of Congress approved July second, eighteen hundred and sixty-two,' the deficiency, if any, in the sum necessary for the support of the said colleges shall be provided for from any moneys in the Treasury not otherwise appropriated.

"Sec. 2. That the Secretary of the Interior is hereby authorized and directed to make examinations and surveys for, and to locate and construct, as herein provided, irrigation works for the storage, diversion, and development of waters, including artesian wells, and to report to Congress at the beginning of each regular session as to the results of such examinations and surveys, giving estimates of cost of all contemplated works, the quantity and location of the lands which can be irrigated therefrom, and all facts relative to the practicability of each irrigation project; also the cost of works in process of construction as well as of those which have been completed.

"Sec. 3. That the Secretary of the Interior shall, before giving the public notice provided for in section four of this Act, withdraw from public entry the lands required for any irrigation works contemplated under the provisions of this Act, and shall restore to public entry any of the lands so withdrawn when, in his judgment, such lands are not required for the purposes of this Act; and the Secretary of the Interior is hereby authorized, at or immediately prior to the time of beginning the surveys for any contemplated irrigation works, to withdraw from entry, except under the homestead laws, any public lands believed to be susceptible of irrigation from said works: *Pro-*

APPENDIX II

vided, That all lands entered and entries made under the homestead laws within areas so withdrawn during such withdrawal shall be subject to all the provisions, limitations, charges, terms, and conditions of this Act; that said surveys shall be prosecuted diligently to completion, and upon the completion thereof, and of the necessary maps, plans, and estimates of cost, the Secretary of the Interior shall determine whether or not said project is practicable and advisable, and if determined to be impracticable or unadvisable he shall thereupon restore said lands to entry; that public lands which it is proposed to irrigate by means of any contemplated works shall be subject to entry only under the provisions of the homestead laws in tracts of not less than forty nor more than one hundred and sixty acres, and shall be subject to the limitations, charges, terms, and conditions herein provided: *Provided*, That the commutation provisions of the homestead laws shall not apply to entries made under this Act.

"Sec. 4. That upon the determination by the Secretary of the Interior that any irrigation project is practicable, he may cause to be let contracts for the construction of the same, in such portions or sections as it may be practicable to construct and complete as parts of the whole project, providing the necessary funds for such portions or sections are available in the reclamation fund, and thereupon he shall give public notice of the lands irrigable under such project, and limit of area per entry, which limit shall represent the acreage which, in the opinion of the Secretary, may be reasonably required for the support of a family upon the lands in question; also of the charges which shall be made per acre upon the said entries, and upon lands in private ownership which may be irrigated by the waters of the said irrigation project, and the number of annual instalments, not exceeding ten, in which such charges shall be paid and the time when such payments shall commence. The said charges shall be determined with a view of returning to the reclamation fund the estimated cost of construction of the project, and shall be apportioned equitably: *Provided*, That in all construction work eight hours shall constitute a day's work, and no Mongolian labor shall be employed thereon.

APPENDIX II

" Sec. 5. That the entryman upon lands to be irrigated by such works shall, in addition to compliance with the homestead laws, reclaim at least one-half of the total irrigable area of his entry for agricultural purposes, and before receiving patent for the lands covered by his entry shall pay to the Government the charges apportioned against such tract, as provided in section four. No right to the use of water for land in private ownership shall be sold for a tract exceeding one hundred and sixty acres to any one landowner, and no such sale shall be made to any landowner unless he be an actual bona fide resident on such land, or occupant thereof residing in the neighborhood of said land, and no such right shall permanently attach until all payments therefor are made. The annual instalments shall be paid to the receiver of the local land office of the district in which the land is situated, and a failure to make any two payments when due shall render the entry subject to cancellation, with the forfeiture of all rights under this Act, as well as of any moneys already paid thereon. All moneys received from the above sources shall be paid into the reclamation fund. Registers and receivers shall be allowed the usual commissions on all moneys paid for lands entered under this Act.

" Sec. 6. That the Secretary of the Interior is hereby authorized and directed to use the reclamation fund for the operation and maintenance of all reservoirs and irrigation works constructed under the provisions of this Act: *Provided*, That when the payments required by this Act are made for the major portion of the lands irrigated from the waters of any of the works herein provided for, then the management and operation of such irrigation works shall pass to the owners of the lands irrigated thereby, to be maintained at their expense under such form of organization and under such rules and regulations as may be acceptable to the Secretary of the Interior: *Provided*, That the title to and the management and operation of the reservoirs and the works necessary for their protection and operation shall remain in the Government until otherwise provided by Congress.

" Sec. 7. That where in carrying out the provisions of this

APPENDIX II

Act it becomes necessary to acquire any rights or property, the Secretary of the Interior is hereby authorized to acquire the same for the United States by purchase or by condemnation under judicial process, and to pay from the reclamation fund the sums which may be needed for that purpose, and it shall be the duty of the Attorney-General of the United States upon every application of the Secretary of the Interior, under this Act, to cause proceedings to be commenced for condemnation within thirty days from the receipt of the application at the Department of Justice.

"Sec. 8. That nothing in this Act shall be construed as affecting or intended to affect or to in any way interfere with the laws of any State or Territory relating to the control, appropriation, use, or distribution of water used in irrigation, or any vested right acquired thereunder, and the Secretary of the Interior, in carrying out the provisions of this Act, shall proceed in conformity with such laws, and nothing herein shall in any way affect any right of any State or of the Federal Government or of any landowner, appropriator, or user of water in, to, or from any interstate stream or the waters thereof: *Provided*, That the right to the use of water acquired under the provisions of this Act shall be appurtenant to the land irrigated, and beneficial use shall be the basis, the measure, and the limit of the right.

"Sec. 9. That it is hereby declared to be the duty of the Secretary of the Interior in carrying out the provisions of this Act, so far as the same may be practicable and subject to the existence of feasible irrigation projects, to expend the major portion of the funds arising from the sale of public lands within each State and Territory hereinbefore named for the benefit of arid and semi-arid lands within the limits of such State or Territory: *Provided*, That the Secretary may temporarily use such portion of said funds for the benefit of arid or semi-arid lands in any particular State or Territory hereinbefore named as he may deem advisable, but when so used the excess shall be restored to the fund as soon as practicable, to the end that ultimately, and in any event, within each ten-year period after the passage of this Act, the expenditures for the benefit of the said States

APPENDIX II

and Territories shall be equalized according to the proportions and subject to the conditions as to practicability and feasibility aforesaid.

"Sec. 10. That the Secretary of the Interior is hereby authorized to perform any and all acts and to make such rules and regulations as may be necessary and proper for the purpose of carrying the provisions of this Act into full force and effect."

INDEX

ADAMS, EDWARD F., quoted, 130.
Africa, 13.
Agricultural industry, persons engaged in, in the United States, 9.
Agua Fria River, Arizona, 249.
Alameda, California, 155.
Alaska, 201.
Anaheim, California, founding and character of, 94.
Ancient canals in Arizona, 35.
Animas River, New Mexico, 239.
Aridity:—Effect of on settlement of Middle West, 17; is key to institutions of West, 30.
Arizona:—The budding civilization of, 247; likeness to region of the Nile, 247; northern part of Territory, 248; Salt River Valley, 248; its irrigation systems, 249; importance of storage plans, 251; Territorial Water-Storage Commission, 252; climate, 252; fruit culture, 253; mineral production, 255; social elements, 256.
Arkansas Valley, Colorado, 166.
Artesia, New Mexico, 243.

Asia Minor, 32.
Austin, Miss, 147.
Australasia, 13.
Aztecs in New Mexico, 34.

BAKERSFIELD, California, 147.
Baldwin, Historian, 34.
Bear Flag, California's day of, 94.
Beaverhead Valley, Montana, 233.
Bee, The Omaha, 266.
Bell, Representative John C., 276.
Berkeley, California, 155.
Billings, Montana, 237.
Bitter Root Valley, Montana, 235.
Blue grass region of Kentucky, compared with arid region, 39.
Boone, Daniel, 15.
Boothe, C. B., 273.
Boyd, David, Historian of Greeley Colony, 88.
Bozeman, Montana, 235.
Brisbane, Albert, 77.
Brodie, Governor Alex. O., quoted, 251.
Brook Farm, 78.
Bullfrog, Nevada, 213.
Butte, Montana, 236.

CACHE LA POUDRE Valley, Colorado, 166.

INDEX

California:—The Empire State of the Pacific, 121; why so little understood, 121; influence of former literature on the subject, 123; speculative tendencies of the past, 128; burdens of fruit-growers before co-operation began, 130; valuable lessons of past twenty years, 131; the State compared with France, 131; agricultural settlements between 1890–1900, 132; profitable lines of production, 133; future of the olive industry, 134; competitors in fruit growing, 135; the mining industry, 136; tendencies of future growth, 137; the coast region, 137; the Santa Clara Valley, 138; Southern California, 139; Sacramento Valley, 141; San Joaquin Valley, 146; birth of raisin industry, 147; effects of fall in price of wheat, 148; possibilities of transportation canals, 149; eastern slope of the Sierra Nevada, 150; Colorado Desert, 151; cities of the State, 154; around San Francisco Bay, 154; in the Sacramento Valley, 156; in the San Joaquin Valley, 156; in the South, 156; See "Evolution of Southern California," 92; orange culture, 100, 148.

Cammas Prairie, Idaho, 195.
Campbell, Douglass, 32.
Carey, Senator J. M., 270.
Carey Act, 193, 228, 270, 323.
Carlsbad, New Mexico, 245.
Carnegie, Andrew, quoted, 9.
Carson Valley, Nevada, 213, 216.

Carthaginians, 34.
Chittenden, Captain Hiram M., 271.
Churchill County, Nevada, 219.
Cimarron River, New Mexico, 240.
City Creek, Salt Lake Valley, Utah, 51.
Clark, Senator William A., 190, 262.
Clark's Fork of the Columbia, 187, 234.
Cleek, Samuel C., 143.
Cody, Col. William F., 227, 322.
Coeur d'Alene Lake, Idaho, 187.
Colonization:—Three great eras of, 12; impulse of American movements, 12; settlement of Atlantic Coast, 14; movement beyond the Alleghanies, 14; settlement of Mississippi Valley, 17; causes of emigration movements, 49.
Colorado:—The New Day in, 150; effects of railroad building, 151; scenery and climate, 153; mineral resources, 154; the Arkansas Valley, 166; the San Luis Valley, 166; the Western Slope, 167; the land of peaches, 169; local patriotism, 171; present economic tendencies, 172.
Colorado Canyon, 248.
Colorado Desert, 151.
Colorado Springs, Colorado, 163.
Columbia River Valley, Washington, 200.
Comstock Lode, production of, 218.
Co-operation:—Influence of aridity in favoring, 31;

INDEX

comparison with conditions in Holland, 32; Utah commercial examples, 64; as employed in the Greeley Colony, Colorado, 89; experience of the Anaheim, California, settlers, 95; how utilized at Riverside, California, 97; California fruit exchanges, 104; necessity of co-operation in Arid America, 328.
Court of Private Claims, 240.
Creoles, French, early settlement of, in Ohio Valley, 15.
Crocker estate, work of, in California, 147.
Cuba, 13.

DAKOTA: growth of, 17; irrigation in, 117.
Damascus, effects of irrigation in, 41.
Davis, Arthur P., 298.
Deccan, lands of the, 36.
Denver, Colorado, 165, 172.
Department of Agriculture, co-operation with Reclamation Service, 302.
"Desert, The," John C. Van Dyke's, quoted, 214.
Douglas County, Nevada, 219.
Drought, the great, of 1890, 265.

EDEN, the garden of, result of irrigation, 42.
Educational advantages in the West, xxiv.
Eggleston, Edward, quoted, 50.
Egypt, 34.
Eight hour day in reclamation work, 296.
Elephant Butte dam site, New Mexico, 240.
Elko County, Nevada, 195, 218.
Ellensburg, Washington, 203.

El Paso, Texas, Herald, quoted, 259.
"Emancipation," x.
Emigration Canyon, Utah, 53.
Esmeralda County, Nevada, 219.
Eureka County, Nevada, 219.

FARMS, statistics of, in the United States, 9.
Fayoom, Province of, 36.
Finney County, Kansas, 109.
Flagg, Jack, in the Rustlers' War, 224.
Flagstaff, Arizona, 248.
Flathead River, Montana, 233.
Fourier, Francois Marie Charles, 77.
Fremont County, Wyoming, 229.
Fulton, Robert L., quoted, 218.

GADSDEN Purchase, 22.
Gallatin Valley, Montana, 233, 235.
Gapen, Dr. Clarke, 340.
Garden City, Kansas, 109, 113.
Gates, George A., xxv.
Geological Survey, organization of, 263.
Gila River, Arizona, 248.
Goldfield, Nevada, 213.
Government reclamation projects:
  Ana River, Oregon, 326.
  Bear Lake, Utah, 326.
  Big Bend, Washington, 327.
  Bellefourche, South Dakota, 321.
  Bismarck, North Dakota, 319.
  Boise-Payette, Idaho, 314.
  Buford-Trenton, North Dakota, 319.
  Chewaucan, Oregon, 326.
  Cheyenne, South Dakota, 326.

INDEX

Dubois, Idaho, 324.
Fort Buford, Montana—North Dakota, 315.
Garden City pumping, Kansas, 324.
Grand Junction, Colorado, 324.
Hondo, New Mexico, 318.
Huntley, Montana, 316.
Klamath, California–Oregon, 311.
Lake De Smet, Wyoming, 327.
La Plata, New Mexico, 325.
Las Vegas, New Mexico, 325.
Malheur, Oregon, 320.
Milk River, Montana, 314.
Minidoka, Idaho, 194, 313.
Niobrara and Snake Rivers, Nebraska, 325.
North Platte, Wyoming–Nebraska, 316.
Okanogan, Washington, 326.
Otter Creek, Oklahoma, 326.
Palouse River, Washington, 322.
Sacramento Valley, California, 144, 324.
Salt River, Arizona, 308, 323.
San Carlos, Arizona, 323.
Shoshone, Wyoming, 228, 322.
Shoshone Indian Reservation, Wyoming, 327.
Silver Creek, Oregon, 327.
Silver Lake, Oregon, 326.
Strawberry Valley, Utah, 326.
Sun River, Montana, 325.
Truckee-Carson, Nevada, 316.
Uncompahgre, Colorado, 312.
Urton Lake, New Mexico, 325.
Utah Lake, Utah, 326.
Walker and Humboldt Rivers, Nevada, 325.
Western Slope of San Diego County, California, 324.
White River, Colorado, 324.
Yuma, California-Arizona, 153, 309.
Grand Junction, Colorado, 168.
Grand River, Colorado, 157.
Great Falls, Montana, 237.
Great Plains, rise of irrigation on, 106.
Greeley, Horace:—Phalanx movement supported by, 77; encourages the Colorado project, 80; his last letter to Meeker, 90.
Greeley Colony, of Colorado:— Its relation to the Phalanx movement of the forties, 77; Meeker proposes the undertaking to Greeley, 80; the colony plan compared with the Fourier ideal, 81; publication of prospectus, 83; irrigation troubles, 84; triumph of the "Greeley potato," 85; social life in the Colony, 87; influence of colony on development of State, 90.
Green River, Colorado, 168.
Green, William Semple, 145; quoted, 146.
Gregory, J. W., 268.
Grunsky, C. E., 299.

HALE, DR. EDWARD EVERETT:— Connection with New Plymouth Colony, of Idaho, xvii, 191; quoted, xvii.
Hall, Benjamin M., 298.
Hall, William Hammond, 145.
Hansbrough, Senator Henry C., 280, 286.
Harrison, Benjamin, quoted, 281.

INDEX

Helena, Montana, 236.
Hilgard, Prof. E. W., quoted, 33, 35, 37.
Hinton, Richard J., 264.
Hitchcock, Secretary E. A., 298.
Holland, compared to Southern California, 92.
Homestead law, effect of on emigration, 17.
Honey Lake Valley, California, 146.
"Horse Heaven" country, Washington, 203.
Hubbard, Elbert, quoted, 262.
Hudson Bay Company, 189.
Humboldt County, Nevada, 219.
Humboldt, Nevada, 216.

IDAHO:—The Crude Strength of 174; contrast between north and south, 174; wonderful water supply, 175; forest area, 176; climate and healthfulness, 188; four periods in history of, 189; area and population, 190; resources and products, 190; prune district, 191; New Plymouth, 191; upper Snake River, 191; private reclamation projects, 193; prices of land, 194; central and northern valleys, 195; the "old-timer," 195.
Illinois, growth of after Revolution, 15.
Incas in South America, 34.
Indiana, growth of after Revolution, 15.
Industrial independence secured by irrigation, 43.
Inyo County, California, 147.
Iowa, growth of, 17.
Irrigation:—Growth of the movement, xi; the miracle of, 41; Damascus the product of, 42; it made the beauties of the Garden of Eden, 42; opposed to land monopoly, 43; as an insurance of crops, 43; unfavorable to employment of servile labor, 44; influence on social life, 45; foundation of scientific agriculture, 47; Mormons, the American pioneers of, 55; comparative cost of private and co-operative systems, 86; district law of California, 148; planks in political platforms, 273; congressional appropriations for, in the West, 275; history of the movement, 361; practical methods of, Appendix I, 333; in humid lands, Appendix I, 333; for national irrigation, National Irrigation Congress, national irrigation law, and National Irrigation Association, see under "N"; for government reclamation projects, see under "G."
Irrigation Age, founding of the, 268.
Irrigation conventions in western Nebraska, 267.

JEFFERSON, President, takes initiative in Western exploration, 23.
Jefferson Valley, Montana, 233.
Johnson County, Wyoming, 226.
Jordan, David Starr; xxv.

KANSAS:—Growth of, 17; irrigation in, 110.
Kentucky, growth of after Revolution, 15.

INDEX

Kennewick, Washington, 203.
King, Clarence, 263.
Kootenai River, Montana, 234.

LA PLATA RIVER, New Mexico, 239.
Lander County, Nevada, 219.
"Lands of the Arid Region," Powell's, 271.
Lassen County, California, 150.
Leland Stanford, junior, university, 156.
Lewis and Clark, the famous journey of, 23.
Libyan Desert, 36.
Lincoln, Abraham, a type of the settlers engaged in a great era of colonization, 17.
Lincoln County, Nevada, 219.
Lippincott, Joseph Barlow, 298; quoted, 145.
Lovelock, Nevada, 216.
Los Angeles, California, 103, 156.
Louisiana Purchase, 22.
Lummis, Charles F., quoted, 157.
Lyon County, Nevada, 219.

MADISON VALLEY, Montana, 233.
Manhattan Valley, Montana, 235.
Manufacturers, persons employed in, in the United States, 9.
Maxwell, George H.: 272, 286; quoted, 279.
Maxwell land grant, New Mexico, 240, 266.
McKinley, William, quoted, 281.
Mead, Elwood, 302.
Meeker, Nathan Cook: experience with Trumbull Phalanx, 79; first trip to the Far West, 80; originates Colorado project, 80; death of, 91.
Mesa City, Arizona, 250.
Milk River Valley, Montana, 233.
Miller, Joaquin, quoted, xii.
Milner, Idaho, 193.
Mining, persons engaged in, in the United States, 10.
Minnesota, growth of, 17.
Missoula, Montana, 237.
Missouri River, in Montana, 233, 236.
Modoc County, California, 150.
Moeris Lake, 36.
Mogollon forest, Arizona, 256.
Mohammedans, their admiration for Damascus, 42.
Mondell, Representative Frank W., 288, 292.
Mongolian labor prohibited in reclamation work, 296.
Monroe, President James, 22.
Montana:—The Prosperity of, 232; influence of mountain topography, 232; early ditches, 233; opportunities for settlers, 234; fruit culture, 235; agricultural college, 235; important valleys, 235; social and political life, 236; cities and towns, 237.
Moors, 34.
Mormons:— their commonwealth, 51; pioneers of American irrigation, 51; they illustrate the natural economic tendencies of irrigation, 52; arrival of first party in Salt Lake Valley, 52; their system of land-ownership, 57; plan of diversified farms, 61; opposed to mining, 63; financial results of their labors for forty years, 67; leading principles

INDEX

of their industrial system, 70; mortgage indebtedness, 71; relation of church organization to industrial success, 74; settlers in San Luis Valley, Colorado, 167; in Uinta Country, Utah, 181.
Musser, A. Milton, 67.

NATIONAL IRRIGATION—Rise of the cause, 261; in what States and Territories, 295; of private lands, explanation of, 305.
National Irrigation Association, 271.
National Irrigation Congresses:
First at Salt Lake, Utah, 1891, 268.
Second at Los Angeles, Cal., 1893, 269.
Third at Denver, Colo., 1894, 270.
Fourth at Albuquerque, N. M., 1895, 270.
Fifth at Phoenix, A. T., 1896, 270.
Sixth at Lincoln, Neb., 1897, 270.
Seventh at Cheyenne, Wyo., 1898, 270.
Eighth at Missoula, Mont., 1899, 270.
Ninth at Chicago, Ill., 1900, 282.
Eleventh at Ogden, Utah 1903, 191.
Twelfth at El Paso, Tex., 1904, 241.
National irrigation law: 276; controversy over authorship of, 287: provisions of, 295; text of Appendix II, 342.
Nebraska: growth of, 17; irrigation conditions in western, 115, 267.

Netherlands, the, civilization of compared with arid region, 32,
Nevada:—The Rising State of, 213; gold fields, 213; popular misconceptions of, 213; oases in, 216; prosperity of farmers, 216; climate and productions, 216; resources of various counties, 218; mineral wealth, 218; future of the State, 220.
New Fork, Wyoming, 229.
New Mexico: The Awakening of, 238; inadequacy of water supplies, 238; the northwestern counties, 239; land grants in, 240; the Pecos Valley, 242; climate and productions, 243; sugar beet culture, 244; pasture lands, 245; Commission of Irrigation, 246; social fabric, 247.
Newell, Frederick Haynes:—on water supply of the plains, 113, 298; quoted, 210, 279.
Newlands, Senator Francis G., 276, 280; quoted, 288.
Newlands irrigation bill, Appendix II, 342.
Nez Perce Indian Reservation, 195.
Nija, Fray Marcos de, 238.
Nile River, silt in, 36.
Ninety-seventh meridan 19, 21.
Nordhoff, Charles, 123, 128.
North, Judge, founder of Riverside Colony, 97.
North Yakima, Washington, 203.
Nye County, Nevada, 219.

OAKLAND, California, 155.
Ohio, growth of after Revolution, 15.
Olive culture, future of, 134.
Ontario, California, 103.

INDEX

Opportunities in the West, xxii.
Oquirrh Mountains, Utah, 53.
Ordinances of 1787, 16.
Oregon:—The State in Transition, 205; transportation facilities, 205; population, 205; bonanza farming, 206; humid and arid sections, 207; need of irrigation, 208; climate and productions, 209, 211; water supplies, 210; central and eastern parts, 210; Lewis and Clark exposition, 212.
Ostrich farm in Arizona, 256.

PALESTINE, 34; resemblance to Great Salt Lake Valley, 53.
Palmer, Gen. William J., 163; 262.
Palo Alto, California, 155.
Palouse country, Idaho, 190.
Parkman Francis, author of *The Oregon Trail*, 23.
Pasco, Washington, 203.
Payette, Idaho, 191.
Pecos Valley, New Mexico, 242.
Pend Oreille Lake, 187.
Phalanx movement, 77.
Phoenix, Arizona, 248.
Pike, Zebulon, 23.
Platte Valley, Colorado, 166.
Plumas County, California, 146.
Plymouth Colony, Idaho, 191.
Pomona, California, 103.
Population of the United States, growth from 1790 to 1900, 10.
Porterville, California, 148.
Portland, Oregon, 209.
Powell Irrigation Survey, 264.
Powell, Major John Wesley, 261, 264, 298.
Prescott, Arizona, 248.
Prescott, Historian, 34.

Prickly Valley, Montana, 233.
Private lands, government irrigation of, 305.
Professions, persons employed in, in the United States, 10.
Prosser, Washington, 203.
Provo, Utah, 177.
Public lands, who may enter, 303.
Puget Sound, Washington, 199.

QUINTON, J. H., 298.

RAILWAY mileage in the United States, 9, 10.
Rain-making experiments, 108.
Raymond, Henry J., debate with Greeley, 79.
Reclamation Service, U. S.: first annual report quoted, 265; board of consulting engineers, 298; organization of, 299; details of work, 303.
Redlands, California, 103.
Reed, Thomas B., quoted 119.
Reid, Whitelaw, quoted, 248.
Reno, Nevada, 213.
"Right Hand of the Continent," Lummis's, 157.
Rio Grande, New Mexico, 240.
Rio Verde, Arizona, 249.
Riverside Colony, California, 97.
Robertson, James, 15.
Robinson, Solon, 79.
Roosevelt, President:—First message, 283; quoted, 284, 285.
Roosevelt, Theodore: 201, 281; quoted, 259.
Rosewater, Edward, 266.
Roswell, New Mexico, 245.
Rustlers' War, 223.

SACRAMENTO, California, 156.

INDEX

Salt Lake City, Utah, plan of, 58.
Salt River, Arizona, 250.
Salt River Valley, Arizona, 248.
San Bernardino Valley, California, 92.
San Diego, California, 156.
San Francisco, California, 154.
San Joaquin Valley, California, 146.
San Juan River, Colorado, 168.
San Juan River, New Mexico, 239.
San Luis Valley, Colorado, 166.
San Pedro, Los Angeles & Salt Lake Railroad (Clark Road), 184, 213.
San Timoteo Hills, California, 92.
Sanders, W. H., 298.
Santa Clara Valley, California, 138.
Savage, H. N., 298.
Schools in the West, xxiv.
Semi-arid region, boundaries of, 109.
Sevier, John, 15.
Shafroth, Representative John F., 275.
Shawhan, Benjamin P., connection with New Plymouth colony, of Idaho, 191.
Sheldon, Lionel A., 269.
Sheridan County, Wyoming, 226.
Shoshone Falls, Idaho, 187.
Smythe, William E.:—connection with the New Plymouth Colony, of Idaho, 191; with rise of the national irrigation cause, 266; founds the *Irrigation Age*, 267.
Snake River, Idaho, 186.
Social life in Arid America, xxi; effect of irrigation on, 46.
Soap, natural, in Nevada, 220.
Soils, effect of aridity on, 37.
South America, 13.
Southern California: evolution of, 92; character and future of, 139.
Spice Islands, 13.
Spokane, Washington, 204.
Stewart, Senator William M., 264.
Stockton, California, 156.
Storey County, Nevada, 219.

TABOR, H. A. W., 172.
Temperaments, eastern and Western, xv.
Tennessee, growth of after Revolution, 15.
Texas, irrigation in, 118.
Thomas, Governor Arthur L., 268.
Tithing-house scrip, Mormon, 63.
Toltecs in Mexico, 34.
Tonopah, Nevada, 213.
Trade and Transportation, persons employed in, in the United States, 10.
Travel, statistics of, 20.
"Triumphant Democracy," Carnegie's, 9.
Truckee, Nevada, 216.
Trumbull Phalanx, of Ohio, 79.
Tucson, Arizona, 248.
Twin Falls Land & Water Co., of Idaho, 193.

UINTA country, Utah, 181.
United States, condition of at close of Revolution, 5.
University of Arizona, experiments in analyzing silt of Colorado River, 40.
University of California, 155.
Utah:—The Pleasant Land of,

INDEX

175; the scene from Capitol Hill, 175; Utah, Salt Lake, and Webber Valleys. 177; mineral resources, 177; climate, 178; agricultural contradictions, 179; lands open to settlement, 180; irrigation laws and customs, 182; construction of Lucin cut-off across Great Salt Lake by Central Pacific Railroad, 184.

VAN DYKE, JOHN C., quoted, 214.
Van Dyke, T. S., quoted, 128.
Vermejo River, New Mexico, 240.

WALCOTT, CHARLES D., Director of Reclamation Service, 298.
Walla Walla, Washington, 203.
Warren, Senator Francis E., 270.
Wasatch Mountains, 53.
Washington:—The Giant, 197; climate and resources, 197, 199, 202; capital for development, 197; strategic situation, 197; population, 198; rainfall, 199, 202; Puget Sound, 199; arid region, 200, 202; important streams, 200; central and eastern part, 200; market for productions, 201.
Washoe, Nevada, 219.
Wealth, total in the United States, 10.

Webber, Thomas G., 65.
Wells, Governor Heber M., 71.
Wenatchee, Washington, 203.
Western Pacific Railroad, 213.
Western Slope of Colorado, 167.
Wheatland Colony, Wyoming, 76.
Wheeler, Benjamin Ide, xxv.
White Pine County, Nevada, 220.
Windmill irrigation in Kansas, 111.
Wisner, G. Y., 298.
Women in the West, xviii.
Woodruff, Wilford, 55.
Worland, Wyoming, 228.
Wyoming:—The Unknown Land of, 221; stock raising industry, 222, 229; Rustlers' War, 223; products and development, 226; Bighorn basin, 227; irrigation development, 228; population, 229; coal, 230; excellence of water laws, 230.
"Wyoming," saddle-horse presented to President Roosevelt, 229.

YAKIMA VALLEY, Washington, 200.
Ybarolla, Senor de, 105.
Yellow River of China, 36.
Young, Brigham, 72, 262.
Yuma, Arizona, 248.

ZION'S Co-operative Mercantile Institution, Utah, 65.

AMERICANA LIBRARY

The City: The Hope of Democracy
By Frederic C. Howe
With a new introduction by Otis A. Pease

Bourbon Democracy of the Middle West, 1865-1896
By Horace Samuel Merrill
With a new introduction by the author

The Deflation of American Ideals: An Ethical Guide for New Dealers
By Edgar Kemler
With a new introduction by Otis L. Graham, Jr.

Borah of Idaho
By Claudius O. Johnson
With a new introduction by the author

The Fight for Conservation
By Gifford Pinchot
With a new introduction by Gerald D. Nash

Upbuilders
By Lincoln Steffens
With a new introduction by Earl Pomeroy

The Progressive Movement
By Benjamin Parke De Witt
With a new introduction by Arthur Mann

Coxey's Army: A Study of the Industrial Army Movement of 1894
By Donald L. McMurry
With a new introduction by John D. Hicks

Jack London and His Times: An Unconventional Biography
By Joan London
With a new introduction by the author

San Francisco's Literary Frontier
By Franklin Walker
With a new introduction by the author

Men of Destiny
By Walter Lippmann
With a new introduction by Richard Lowitt

*Woman Suffrage and Politics:
The Inner Story of the Suffrage Movement*
By Carrie Chapman Catt and Nettie H. Shuler
With a new introduction by T. A. Larson

The Dry Decade
By Charles Merz
With a new introduction by the author

The Conquest of Arid America
By William E. Smythe
With a new introduction by Lawrence B. Lee

*The Territories and the United States, 1861–1890:
Studies in Colonial Administration*
By Earl S. Pomeroy
With a new introduction by the author

DATE DUE

WITHDRAWN
from
Funderburg Library

FUNDERBURG LIBRARY
MANCHESTER COLLEGE

917.8
Sm99c